# Architecture and Design of the Linux Storage Stack

Gain a deep understanding of the Linux storage landscape and its well-coordinated layers

**Muhammad Umer**

BIRMINGHAM—MUMBAI

# Architecture and Design of the Linux Storage Stack

**Group Product Manager**: Pavan Ramchandani

**Publishing Product Manager**: Prachi Sawant

**Senior Editor**: Runcil Rebello

**Technical Editor**: Arjun Varma

**Copy Editor**: Safis Editing

**Project Coordinator**: Ashwin Kharwa

**Proofreader**: Safis Editing

**Indexer**: Pratik Shirodkar

**Production Designer**: Alishon Mendonca

**Marketing Coordinator**: Marylou De Mello

First published: July 2023

Production reference: 280623

Published by Packt Publishing Ltd.

Livery Place

35 Livery Street

Birmingham

B3 2PB, UK.

ISBN 978-1-83763-996-0

www.packtpub.com

*To Ammi, Abu, Hammad, Sharjeel, and Amna, who faced some arduous times because of me and taught me the meaning of sacrifice, patience, hard work, and perseverance. You have always stood by my side on good days and stood even closer on my bad days.*

*And to Ahad, who brings joy and wonder into our lives.*

*I dedicate this book to you with all my love.*

# Contributors

## About the author

**Muhammad Umer** is a systems engineer and a trainer, with more than six years of experience working with Linux-based systems, **high availability** (HA) design and architecture, tuning operating systems, and underlying hardware for optimal performance and root cause analysis. He likes to keep himself updated with the latest happenings in the world of technology and is a **Red Hat Certified Architect** (**RHCA**). Umer is an avid Linux enthusiast, with a strong focus on storage-related technologies. This passion has driven him to author a book on the topic, drawing on his extensive experience in the field.

*I want to thank the people with whom I've worked over the years. Each one of them has taught me something in one way or another. I'm also grateful to the wonderful team at Packt for their support throughout this journey.*

# About the reviewer

**Pradeep Roy Kandru** is a Linux kernel programmer and has extensive experience working in different layers of the Kernel. Having worked in both technical and development roles, his mainstay has been a broad base of skills and the ability to work on diverse projects. Academically, he has secured All India Rank 43 in the Graduate Aptitude Test in Engineering and holds a master's degree in internet science engineering, gained at the Department of Computer Science at the **Indian Institute of Science (IISc)**, Bangalore, India. Along with reviewing other technical books, he also works on his own forthcoming technical book as an author. He worked for the companies IBM, India Software Labs, Veritas, and Depik Technologies. His current interest includes eBPF observability.

# Table of Contents

# 3

# Exploring the Actual Filesystems Under the VFS                    53

# Part 2: Navigating Through the Block Layer

# 4

# Understanding the Block Layer, Block Devices, and Data Structures    83

# 5

# Understanding the Block Layer, Multi-Queue, and Device Mapper    101

# 6

# Understanding I/O Handling and Scheduling in the Block Layer    117

# Part 3: Descending into the Physical Layer

# 7

# The SCSI Subsystem    139

# 8

## Illustrating the Layout of Physical Media    153

# Part 4: Analyzing and Troubleshooting Storage Performance

# 9

## Analyzing Physical Storage Performance    171

# 10

## Analyzing Filesystems and the Block Layer                    187

# 11

## Tuning the I/O Stack                                         207

# Preface

The major benefit of open source operating systems such as Linux is that anyone can dig deep and uncover how they work under the hood. Even with the astronomical advances in software development, the Linux kernel remains one of the most complex pieces of code. Developers, programmers, and would-be kernel hackers are constantly diving into the kernel code and are pushing for new features. Hobbyists and enthusiasts like me are trying to understand and unravel mysteries.

As one of those enthusiasts, I have spent quite some time exploring the intricacies of the Linux storage stack. From simple hard drives to complex networked storage systems, Linux is at the heart of many of the world's most complex storage technologies. Over the last few years, I've had a chance to work on Linux and several storage technologies, which piqued my interest in this area. This book is the result of that exploration. I've tried peeling off all the layers of the Linux storage stack to show how they all work in unison. My goal is to share what I have learned with others who share my fascination with this topic.

This book provides an in-depth and conceptual overview of the Linux storage stack. It covers all of the major components of the stack and includes an in-depth and detailed analysis of the storage subsystem and its architecture, the virtual filesystem layer, different filesystems and the differences in their implementations, the block layer, multi-queue and device mapper frameworks, scheduling, and physical layers. It also covers a variety of topics relating to storage performance analysis, tuning, and troubleshooting.

I'm sure anyone who wants to widen their understanding of Linux and its storage landscape will find this book informative and useful.

## Who this book is for

The primary target audience for this book is Linux system and storage administrators, engineers, Linux professionals, the Linux community in general, and anyone who wants to widen their understanding of Linux. When working in any environment, it is imperative to have a deep understanding of the technology that you're working on. I believe this book will give you a better understanding of Linux internals, equip you with the necessary knowledge, and increase your general interest in the subject.

# What this book covers

*Chapter 1, Where It All Starts From – The Virtual Filesystem*, provides an introduction to the **virtual filesystem** (**VFS**) in the Linux kernel. The chapter will explain the pivotal role of the VFS in the I/O stack and provide a strong conceptual understanding of VFS as it is the starting point of an I/O request in Linux.

*Chapter 2, Explaining the Data Structures in a VFS*, introduces the various data structures utilized by the VFS in the kernel. The usage of structures by the kernel such as inodes, directory entries, and file objects to store file metadata, directories, and open files will be explained. Furthermore, the method by which the superblock structure enables the kernel to record filesystem characteristics will also be covered. Finally, the page caching mechanism in the kernel will be explained.

*Chapter 3, Exploring the Actual Filesystems Under the VFS*, introduces the concept of filesystems in Linux. One of the most popular block-based filesystems in Linux, the extended filesystem, will be explained. Additionally, important filesystem concepts such as journaling and copy-on-write will be discussed in detail. The chapter will also delve into the differences between file and block I/O and explore the network filesystem. Finally, the concept of userspace filesystems will be introduced.

*Chapter 4, Understanding the Block Layer, Block Devices, and Data Structures*, introduces the block layer in the kernel. The chapter will explain the concept of block devices and how they differ from character devices, and covers the major data structures in the block layer.

*Chapter 5, Understanding the Block Layer, Multi-Queue, and Device Mapper*, introduces the device mapper framework and the multi-queue block I/O queuing mechanism in the kernel. The chapter will explain how the multi-queue framework improves the performance of I/O operations on modern storage devices. Further, the role of the device mapper in laying the foundation for features such as LVM will be explained.

*Chapter 6, Understanding I/O Handling and Scheduling in the Block Layer*, discusses the different I/O handling mechanisms and I/O schedulers in the kernel. Operations such as merging, coalescing, and plugging are discussed in terms of how they are handled in the block layer. The different I/O schedulers supported in the Linux kernel are also explained in detail, along with the differences in their operational logic.

*Chapter 7, The SCSI Subsystem*, focuses on the SCSI subsystem in Linux and its multi-layered architecture. The chapter will explain the multilayered SCSI architecture, the SCSI device addressing mechanism, and the main data structures in the SCSI layer.

*Chapter 8, Illustrating the Layout of Physical Media*, discusses the different storage media available today, and the differences in their architecture. The traditional mechanical drives, solid-state drives, and the new NVMe interface are compared.

*Chapter 9, Analyzing Physical Storage Performance*, covers the performance analysis and characteristics of the storage subsystem. This chapter will present the different metrics that can be used to assess the performance of physical storage. Later on, the different tools and mechanisms that can be used to gauge storage performance are discussed.

*Chapter 10, Analyzing Filesystems and the Block Layer,* focuses on the techniques that can be used to analyze the performance of the block layer and filesystems. This chapter will explain the different types of filesystem I/O and the factors that can affect the I/O request of an application. Later on, the different tools and tracing mechanisms, such as the Berkeley Packet Filter compiler collection, are looked at to identify potential bottlenecks at each layer.

*Chapter 11, Tuning the I/O Stack,* discusses some recommended practices for tuning the underlying storage layer for application needs. The different options that can be used to tune the performance at each layer are discussed.

## To get the most out of this book

The main objective of the book is to comprehend the inner workings of the Linux kernel and its major subsystems. Therefore, to get the most out of this book you'll need to have a decent understanding of operating system concepts in general and Linux in particular. Above all, it is important to approach these topics with patience, curiosity, and a willingness to learn.

| Software/hardware covered in the book | Operating system requirements |
|---|---|
| | Linux |

The commands for installing specific packages are included in the technical requirements section of each chapter.

## Download the example code files

We have code bundles from our rich catalog of books and videos available at `https://github.com/PacktPublishing/`. Check them out!

## Download the color images

We also provide a PDF file that has color images of the screenshots and diagrams used in this book. You can download it here: `https://packt.link/Uz9ge`

## Conventions used

There are a number of text conventions used throughout this book.

`Code in text`: Indicates code words in text, database table names, folder names, filenames, file extensions, pathnames, dummy URLs, user input, and Twitter handles. Here is an example: "If we look at the `sd*` devices in `/dev`, notice that the file type is being shown as b, for block devices."

A block of code is set as follows:

```
struct block_device {
        sector_t                bd_start_sect;
        sector_t                bd_nr_sectors;
        struct disk_stats __percpu *bd_stats;
        unsigned long           bd_stamp;
        bool                    bd_read_only;
        dev_t                   bd_dev;
        atomic_t                bd_openers;
        struct inode *          bd_inode;
[........]
```

When we wish to draw your attention to a particular part of a code block, the relevant lines or items are set in bold:

```
struct block_device {
        sector_t                bd_start_sect;
        sector_t                bd_nr_sectors;
        struct disk_stats __percpu *bd_stats;
        unsigned long           bd_stamp;
        bool                    bd_read_only;
        dev_t                   bd_dev;
        atomic_t                bd_openers;
        struct inode *          bd_inode;
[........]
```

Any command-line input or output is written as follows:

```
[root@linuxbox ~]# find / -inum 67118958 -exec ls -l {} \;
-rw-r--r-- 1 root root 220 Jun 15 22:30 /etc/hosts
[root@linuxbox ~]#
```

**Bold**: Indicates a new term, an important word, or words that you see onscreen. For instance, words in menus or dialog boxes appear in **bold**. Here is an example: "In the case of a directory, the **type** field in an inode is a **directory**."

---

**Tips or important notes**
Appear like this.

---

# Get in touch

Feedback from our readers is always welcome.

**General feedback**: If you have questions about any aspect of this book, email us at `customercare@packtpub.com` and mention the book title in the subject of your message.

**Errata**: Although we have taken every care to ensure the accuracy of our content, mistakes do happen. If you have found a mistake in this book, we would be grateful if you would report this to us. Please visit `www.packtpub.com/support/errata` and fill in the form.

**Piracy**: If you come across any illegal copies of our works in any form on the internet, we would be grateful if you would provide us with the location address or website name. Please contact us at `copyright@packt.com` with a link to the material.

**If you are interested in becoming an author**: If there is a topic that you have expertise in and you are interested in either writing or contributing to a book, please visit `authors.packtpub.com`.

# Share Your Thoughts

Once you've read *Architecture and Design of the Linux Storage Stack*, we'd love to hear your thoughts! Scan the QR code below to go straight to the Amazon review page for this book and share your feedback.

`https://packt.link/r/1837639965`

Your review is important to us and the tech community and will help us make sure we're delivering excellent quality content.

# Download a free PDF copy of this book

Thanks for purchasing this book!

Do you like to read on the go but are unable to carry your print books everywhere?

Is your eBook purchase not compatible with the device of your choice?

Don't worry, now with every Packt book you get a DRM-free PDF version of that book at no cost.

Read anywhere, any place, on any device. Search, copy, and paste code from your favorite technical books directly into your application.

The perks don't stop there, you can get exclusive access to discounts, newsletters, and great free content in your inbox daily

Follow these simple steps to get the benefits:

1.  Scan the QR code or visit the link below

https://packt.link/free-ebook/9781837639960

2.  Submit your proof of purchase
3.  That's it! We'll send your free PDF and other benefits to your email directly

# Part 1: Diving into the Virtual Filesystem

This part offers a detailed introduction to the **virtual filesystem** (**VFS**) layer and the actual filesystems underneath it. You will learn about VFS, its major data structures, the extended filesystem family, and the major concepts associated with the different filesystems in Linux.

This part contains the following chapters:

- *Chapter 1, Where It All Starts From – The Virtual Filesystem*
- *Chapter 2, Explaining the Data Structures in a VFS*
- *Chapter 3, Exploring the Actual Filesystems Under the VFS*

# 1

# Where It All Starts From – The Virtual Filesystem

Even with astronomical advances in software development, the Linux kernel remains one of the most complex pieces of code. Developers, programmers, and would-be kernel hackers constantly look to dive into kernel code and push for new features, whereas hobbyists and enthusiasts try to understand and unravel those mysteries.

Naturally, a lot has been written on Linux and its internal workings, from general administration to kernel programming. Over the decades, hundreds of books have been published, which cover a diverse range of important operating system topics, such as process creation, threading, memory management, virtualization, filesystem implementations, and CPU scheduling. This book that you've picked up (thank you!) will focus on the storage stack in Linux and its multilayered organization.

We'll start by introducing the Virtual Filesystem in the Linux kernel and its pivotal role in allowing end user programs to access data on filesystems. Since we intend to cover the entire storage stack in this book, from top to bottom, getting a deeper understanding of the Virtual Filesystem is extremely important, as it is the starting point of an I/O request in the kernel. We'll introduce the concept of user space and kernel space, understand system calls, and see how the *Everything is a file* philosophy in Linux is tied to the Virtual Filesystem.

In this chapter, we're going to cover the following main topics:

- Understanding storage in a modern-day data center
- Defining system calls
- Explaining the need for a Virtual Filesystem
- Describing the Virtual Filesystem
- Explaining the *Everything is a file* philosophy

## Technical requirements

Before going any further, I think is important to acknowledge here that certain technical topics may be more challenging for beginners to comprehend than others. Since the goal here is to comprehend the inner workings of the Linux kernel and its major subsystems, it will be helpful to have a decent foundational understanding of operating system concepts in general and Linux in particular. Above all, it is important to approach these topics with patience, curiosity, and a willingness to learn.

The commands and examples presented in this chapter are distribution-agnostic and can be run on any Linux operating system, such as Debian, Ubuntu, Red Hat, and Fedora. There are a few references to the kernel source code. If you want to download the kernel source, you can download it from https://www.kernel.org. The operating system packages relevant to this chapter can be installed as follows:

- For Ubuntu/Debian:

  - `sudo apt install strace`

  - `sudo apt install bcc`

- For Fedora/CentOS/Red Hat-based systems:

  - `sudo yum install strace`

  - `sudo yum install bcc-tools`

## Understanding storage in a modern-day data center

*It is a capital mistake to theorize before one has data. Insensibly one begins to twist facts to suit theories, instead of theories to suit facts. – Sir Arthur Conan Doyle*

Compute, storage, and networking are the basic building blocks of any infrastructure. How well your applications do is often dependent on the combined performance of these three layers. The workloads running in a modern data center vary from streaming services to machine learning applications. With the meteoric rise and adoption of cloud computing platforms, all the basic building blocks are now abstracted from the end user. Adding more hardware resources to your application, as it becomes resource-hungry, is the new normal. Troubleshooting performance issues is often skipped in favor of migrating applications to better hardware platforms.

Of the three building blocks, compute, storage, and networking, storage is often considered the bottleneck in most scenarios. For applications such as databases, the performance of the underlying storage is of prime importance. In cases where infrastructure hosts mission-critical and time-sensitive applications such as **Online Transaction Processing (OLTP)**, the performance of storage frequently comes under the radar. The smallest of delays in servicing I/O requests can impact the overall response of the application.

The most common metric used to measure storage performance is latency. The response times of storage devices are usually measured in milliseconds. Compare that with your average processor or memory, where such measurements are measured in nanoseconds, and you'll see how the performance of the storage layer can impact the overall working of your system. This results in a state of incongruity between the application requirements and what the underlying storage can actually deliver. For the last few years, most of the advancements in modern-day storage drives have been geared toward sizing – the **capacity arena**. However, performance improvement of the storage hardware has not progressed at the same rate. Compared to the compute functions, the performance of storage pales in comparison. For these reasons, it is often termed the *three-legged dog of the data center*.

Having made a point about the choice of a storage medium, it's pertinent to note that no matter how powerful it is, the hardware will always have limitations in its functionality. It's equally important for the application and operating system to tune themselves according to the hardware. Fine-tuning your application, operating system, and filesystem parameters can give a major boost to the overall performance. To utilize the underlying hardware to its full potential, all layers of the I/O hierarchy need to function efficiently.

## Interacting with storage in Linux

The Linux kernel makes a clear distinction between the user space and kernel space processes. All the hardware resources, such as CPU, memory, and storage, lie in the kernel space. For any user space application wanting to access the resources in kernel space, it has to generate a **system call**, as shown in *Figure 1.1*:

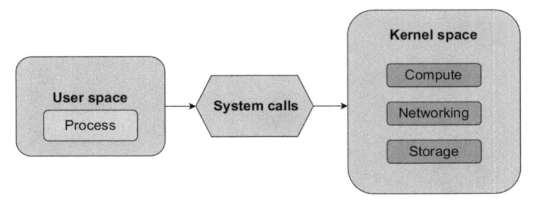

Figure 1.1 – The interaction between user space and kernel space

**User space** refers to all the applications and processes that live outside of the kernel. The kernel space includes programs such as device drivers, which have unrestricted access to the underlying hardware. The user space can be considered a form of sandboxing to restrict the end user programs from modifying critical kernel functions.

This concept of user and kernel space is deeply rooted in the design of modern processors. A traditional x86 CPU uses the concept of protection domains, called **rings**, to share and limit access to hardware resources. Processors offer four rings or modes, which are numbered from 0 to 3. Modern-day processors are designed to operate in two of these modes, ring 0 and ring 3. The user space applications are handled in ring 3, which has limited access to kernel resources. The kernel occupies ring 0. This is where the kernel code executes and interacts with the underlying hardware resources.

When processes need to read from or write to a file, they need to interact with the filesystem structures on top of the physical disk. Every filesystem uses different methods to organize data on the physical disk. The request from the process doesn't directly reach the filesystem or physical disk. In order for the I/O request of the process to be served by the physical disk, it has to traverse through the entire storage hierarchy in the kernel. The first layer in that hierarchy is known as the **Virtual Filesystem**. The following figure, *Figure 1.2*, highlights the major components of the Virtual Filesystem:

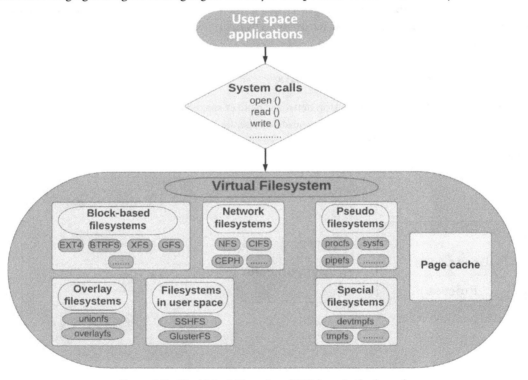

Figure 1.2 – The Virtual Filesystem (VFS) layer in the kernel

The storage stack in Linux consists of a multitude of cohesive layers, all of which ensure that the access to physical storage media is abstracted through a unified interface. As we move forward, we're going to build upon this structure and add more layers. We'll try to dig deep into each of them and see how they all work in harmony.

This chapter will focus solely on the Virtual Filesystem and its various features. In the coming chapters, we're going to explain and uncover some under-the-hood workings of the more frequently used filesystems in Linux. However, bearing in mind the number of times the word *filesystem* is going to be used here, I think it's prudent to briefly categorize the different filesystem types, just to avoid any confusion:

- **Block filesystems**: Block- or disk-based filesystems are the most common way to store user data. As a regular operating system user, these are the filesystems that users mostly interact with. Filesystems such as **Extended filesystem version 2/3/4 (Ext 2/3/4)**, **Extent filesystem (XFS)**, Btrfs, FAT, and NTFS are all categorized as disk-based or block filesystems. These filesystems speak in terms of **blocks**. The block size is a property of the filesystem, and it can only be set when creating a filesystem on a device. The block size indicates what size the filesystem will use when reading or writing data. We can refer to it as the logical unit of storage allocation and retrieval for a filesystem. A device that can be accessed in terms of blocks is, therefore, called a **block device**. Any storage device attached to a computer, whether it is a hard drive or an external USB, can be classified as a block device. Traditionally, block filesystems are mounted on a single host and do not allow sharing between multiple hosts.

- **Clustered filesystems**: Clustered filesystems are also block filesystems and use block-based access methods to read and write data. The difference is that they allow a single filesystem to be mounted and used simultaneously by multiple hosts. Clustered filesystems are based on the concept of **shared storage**, meaning that multiple hosts can concurrently access the same block device. Common clustered filesystems used in Linux are Red Hat's **Global File System 2 (GFS2)** and **Oracle Clustered File System (OCFS)**.

- **Network filesystems (NFS)**: NFS is a protocol that allows for remote file sharing. Unlike regular block filesystems, NFS is based on the concept of sharing data between multiple hosts. NFS works with the concept of a client and a server. The backend storage is provided by an NFS server. The host systems on which the NFS filesystem is mounted are called **clients**. The connectivity between the client and server is achieved using conventional Ethernet. All NFS clients share a single copy of the file on the NFS server. NFS doesn't offer the same performance as block filesystems, but it is still used in enterprise environments, mostly to store long-term backups and share common data.

- **Pseudo filesystems**: Pseudo filesystems exist in the kernel and generate their content dynamically. They are not used to store data persistently. They do not behave like regular disk-based filesystems such as Ext4 or XFS. The main purpose of a pseudo filesystem is to allow the user space programs to interact with the kernel. Directories such as `/proc` (procfs) and `/sys` (sysfs) fall under this category. These directories contain virtual or temporary files, which include information about the different kernel subsystems. These pseudo filesystems are also a part of the Virtual Filesystem landscape, as we'll see in the *Everything is a file* section.

Now that we have a basic idea about user space, kernel space, and the different types of filesystems, let's explain how an application can request resources in kernel space through system calls.

# Understanding system calls

While looking at the figure explaining the interaction between applications and the Virtual Filesystem, you may have noticed the intermediary layer between user space programs and the Virtual Filesystem; that layer is known as the **system call interface**. To request some service from the kernel, user space programs invoke the system call interface. These system calls provide the means for end user applications to access the resources in the kernel space, such as the processor, memory, and storage. The system call interface serves three main purposes:

- **Ensuring security**: System calls prevent user space applications from directly modifying resources in the kernel space

- **Abstraction**: Applications do not need to concern themselves with the underlying hardware specifications

- **Portability**: User programs can be run correctly on all kernels that implement the same set of interfaces

There's often some confusion about the differences between system calls and an **application programming interface** (**API**). An API is a set of programming interfaces used by a program. These interfaces define a method of communication between two components. An API is implemented in user space and outlines how to acquire a particular service. A system call is a much lower-level mechanism that uses interrupts to make an explicit request to the kernel. The system call interface is provided by the standard C library in Linux.

If the system call generated by the calling process succeeds, a file descriptor is returned. A **file descriptor** is an integer number that is used to access files. For example, when a file is opened using the open () system call, a file descriptor is returned to the calling process. Once a file has been opened, programs use the file descriptor to perform operations on the file. All read, write, and other operations are performed using the file descriptor.

Every process always has a minimum of three files opened – standard input, standard output, and standard error – represented by the 0, 1, and 2 file descriptors, respectively. The next file opened will be assigned the file descriptor value of 3. If we do some file listing through ls and run a simple strace, the open system call will return a value of 3, which is the file descriptor representing the file – /etc/hosts, in this case. After that, this file descriptor value of 3 is used by the fstat and close calls to perform further operations:

```
strace ls /etc/hosts
root@linuxbox:~# strace ls /etc/hosts
execve("/bin/ls", ["ls", "/etc/hosts"], 0x7ffdee289b48 /* 22 vars */)
= 0
brk(NULL)                            = 0x562b97fc6000
access("/etc/ld.so.nohwcap", F_OK)   = -1 ENOENT (No such file or
directory)
access("/etc/ld.so.preload", R_OK)   = -1 ENOENT (No such file or
```

```
directory)
openat(AT_FDCWD, "/etc/ld.so.cache", O_RDONLY|O_CLOEXEC) = 3
fstat(3, {st_mode=S_IFREG|0644, st_size=140454, ...}) = 0
mmap(NULL, 140454, PROT_READ, MAP_PRIVATE, 3, 0) = 0x7fbaa2519000
close(3)                              = 0
access("/etc/ld.so.nohwcap", F_OK)    = -1 ENOENT (No such file or
directory)
openat(AT_FDCWD, "/lib/x86_64-linux-gnu/libselinux.so.1", O_RDONLY|O_
CLOEXEC) = 3
```

[The rest of the code is skipped for brevity.]

On x86 systems, there are around 330 system calls. This number could be different for other architectures. Each system call is represented by a unique integer number. You can list the available system calls on your system using the `ausyscall` command. This will list the system calls and their corresponding integer values:

```
ausyscall -dump
root@linuxbox:~# ausyscall --dump
Using x86_64 syscall table:
0        read
1        write
2        open
3        close
4        stat
5        fstat
6        lstat
7        poll
8        lseek
9        mmap
10       mprotect
```

[The rest of the code is skipped for brevity.]

```
root@linuxbox:~# ausyscall --dump|wc -l
334
root@linuxbox:~#
```

The following table lists some common system calls:

| System call | Description |
| --- | --- |
| `open ()`, `close ()` | Open and close files |
| `create ()` | Create a file |
| `chroot ()` | Change the `root` directory |
| `mount ()`, `umount ()` | Mount and unmount filesystems |
| `lseek ()` | Change the pointer position in a file |
| `read ()`, `write ()` | Read and write in a file |
| `stat ()`, `fstat ()` | Get a file status |
| `statfs ()`, `fstatfs ()` | Get filesystem statistics |
| `execve ()` | Execute the program referred to by pathname |
| `access ()` | Checks whether the calling process can access the file pathname |
| `mmap ()` | Creates a new mapping in the virtual address space of the calling process |

Table 1.1 – Some common system calls

So, what role do the system calls play in interacting with filesystems? As we'll see in the succeeding section, when a user space process generates a system call to access resources in the kernel space, the first component it interacts with is the Virtual Filesystem. This system call is first handled by the corresponding system call handler in the kernel, and after validating the operation requested, the handler makes a call to the appropriate function in the VFS layer. The VFS layer passes the request on to the appropriate filesystem driver module, which performs the actual operations on the file.

We need to understand the *why* here – why would the process interact with the Virtual Filesystem and not the actual filesystem on the disk? In the upcoming section, we'll try to figure this out.

To summarize, the system calls interface in Linux implements generic methods that can be used by the applications in user space to access resources in the kernel space.

## Explaining the need for a Virtual Filesystem

A **standard filesystem** is a set of data structures that determine how user data is organized on a disk. End users are able to interact with this standard filesystem through regular file access methods and perform common tasks. Every operating system (Linux or non-Linux) provides at least one such filesystem, and naturally, each of them claims to be *better, faster, and more secure* than the other. A great majority of modern Linux distributions use **XFS or Ext4** as the default filesystem. These filesystems have several features and are considered stable and reliable for daily usage.

However, the support for filesystems in Linux is not limited to only these two. One of the great benefits of using Linux is that it offers support for multiple filesystems, all of which can be considered perfectly acceptable alternatives to Ext4 and XFS. Because of this, Linux can peacefully coexist with other operating systems. Some of the more commonly used filesystems include older versions of Ext4, such as Ext2 and Ext3, Btrfs, ReiserFS, OpenZFS, FAT, and NTFS. When using multiple partitions, users can choose from a long list of available filesystems and create a different one on every disk partition as per their needs.

The smallest addressable unit of a physical hard drive is a sector. For filesystems, the smallest writable unit is called a block. A **block** can be considered a group of consecutive sectors. All operations by a filesystem are performed in terms of blocks. There is no singular way in which these blocks are addressed and organized by different filesystems. Each filesystem may use a different set of data structures to allocate and store data on these blocks. The presence of a different filesystem on each storage partition can be difficult to manage. Given the wide range of supported filesystems in Linux, imagine if applications needed to understand the distinct details of every filesystem. In order to be compatible with a filesystem, the application would need to implement a unique access method for each filesystem it uses. This would make the design of an application almost impractical.

Abstraction interfaces play a critical role in the Linux kernel. In Linux, regardless of the filesystem being used, the end users or applications can interact with the filesystem using uniform access methods. All this is achieved through the Virtual Filesystem layer, which hides the filesystem implementations under an all-inclusive interface.

## Describing the VFS

To ensure that applications do not face any such obstacles (as mentioned earlier) when working with different filesystems, the Linux kernel implements a layer between end user applications and the filesystem on which data is being stored. This layer is known as the **Virtual Filesystem** (**VFS**). The VFS is not a standard filesystem, such as Ext4 or XFS. (There is no mkfs.vfs command!) For this reason, some prefer the term **Virtual Filesystem Switch**.

Think of the *magic wardrobe* from *The Chronicles of Narnia*. The wardrobe is actually a portal to the magical world of Narnia. Once you step through the wardrobe, you can explore the new world and interact with its inhabitants. The wardrobe facilitates accessing the magical world. In a similar way, the VFS provides a doorway to different filesystems.

The VFS defines a generic interface that allows multiple filesystems to coexist in Linux. It's worth mentioning again that with the VFS, we're not talking about a standard block-based filesystem. We're talking about an abstraction layer that provides a link between the end user application and the actual block filesystems. Through the standardization implemented in the VFS, applications can perform read and write operations, without worrying about the underlying filesystem.

As shown in *Figure 1.3*, the VFS is interposed between the user space programs and actual filesystems:

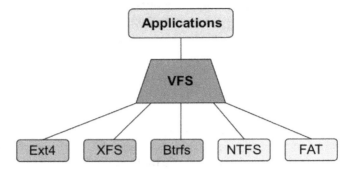

Figure 1.3 – The VFS acts as a bridge between user space programs and filesystems

For the VFS to provide services to both parties, the following has to apply:

- All end user applications need to define their filesystem operations in terms of the standard interface provided by the VFS

- Every filesystem needs to provide an implementation of the common interface provided by the VFS

We explained that applications in user space need to generate system calls when they want to access resources in the kernel space. Through the abstraction provided by the VFS, system calls such as read() and write() function properly, regardless of the filesystem in use. These system calls work across filesystem boundaries. We don't need a special mechanism to move data to a different or non-native filesystem. For instance, we can easily move data from an Ext4 filesystem to XFS, and vice versa. At a very high level, when a process issues the read() or write() system call to read or write a file, the VFS will search for the filesystem driver to use and forward these system calls to that driver.

## Implementing a common filesystem interface through the VFS

The primary goal of the VFS is to represent a diverse set of filesystems in the kernel with minimum overhead. When a process requests a read or write operation on a file, the kernel substitutes this with the filesystem-specific function on which the file resides. In order to achieve this, every filesystem must adapt itself in terms of the VFS.

Let's go through the following example for a better understanding.

Consider the example of the cp (copy) command in Linux. Let's suppose we're trying to copy a file from an Ext4 to an XFS filesystem. How does this copy operation complete? How does the cp command interact with the two filesystems? Have a look at *Figure 1.4*:

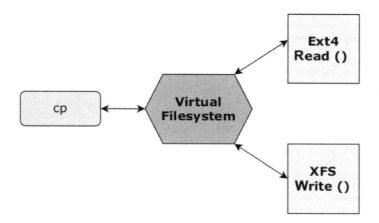

Figure 1.4 – The VFS ensures interoperability between different filesystems

First off, the `cp` command doesn't care about the filesystems being used. We've defined the VFS as the layer that implements abstraction. So, the `cp` command doesn't need to concern itself about the filesystem details. It will interact with the VFS layer through the standard system call interface. Specifically, it will issue the `open ()` and `read ()` system calls to open and read the file to be copied. An open file is represented by the file data structure in the kernel (as we'll learn in the next chapter, *Chapter 2, Explaining the Data Structures in a VFS*.

When `cp` generates these generic system calls, the kernel will redirect these calls to the appropriate function of the filesystem through a pointer, on which the file resides. To copy the file to the XFS filesystem, the `write ()` system call is passed to the VFS. This will again be redirected to the particular function of the XFS filesystem that implements this feature. Through system calls issued to the VFS, the `cp` process can perform a copy operation using the `read ()` method of Ext4 and the `write ()` method of XFS. Just like a *switch*, the VFS will switch the common file access methods between their designated filesystem implementations.

The read, write, or any other function for that matter does not have a default definition in the kernel – hence the name **virtual**. The interpretation of these operations depends upon the underlying filesystem. Just like user programs that take advantage of this abstraction offered by the VFS, filesystems also reap the benefits of this approach. Common access methods for files do not need to be reimplemented by filesystems.

That was pretty neat, right? But what if we want to copy something from Ext4 to a non-native filesystem? Filesystems such as Ext4, XFS, and Btrfs were specifically designed for Linux. What if one of the filesystems involved in this operation is FAT or NTFS?

Admittedly, the design of the VFS is biased toward filesystems that come from the Linux tribe. To an end user, there is a clear distinction between a file and a directory. In the Linux philosophy, everything is a file, including directories. Filesystems native to Linux, such as Ext4 and XFS, were designed keeping these nuances in mind. Because of the differences in the implementation, non-native filesystems such as FAT and NTFS do not support all of the VFS operations. The VFS in Linux uses structures such as inodes, superblocks, and directory entries to represent a generic view of a filesystem. Non-native Linux filesystems do not speak in terms of these structures. So how does Linux accommodate these filesystems? Take the example of the FAT filesystem. The FAT filesystem comes from a different world and doesn't use these structures to represent files and directories. It doesn't treat directories as files. So, how does the VFS interact with the FAT filesystem?

All filesystem-related operations in the kernel are firmly integrated with the VFS data structures. To accommodate non-native filesystems on Linux, the kernel constructs the corresponding data structures dynamically. For instance, to satisfy the common file model for filesystems such as FAT, files corresponding to directories will be created in memory on the fly. These *files* are *virtual* and will only exist in memory. This is an important concept to understand. On native filesystems, structures such as inodes and superblocks are not only present in memory but also stored on the physical medium itself. Conversely, non-Linux filesystems merely have to perform the enactment of such structures in memory.

## Peeking at the source code

If we take a look at the kernel source code, the different functions provided by the VFS are present in the fs directory. All source files ending in .c contain implementations of the different VFS methods. The subdirectories contain specific filesystem implementations, as shown in *Figure 1.5*:

index : kernel/git/stable/linux.git

Linux kernel stable tree

about    summary    refs    log    **tree**    commit    diff    stats

path: root/fs

| Mode | Name | Size |
|------|------|------|
| d--------- | 9p | 724 |
| -rw-r--r-- | Kconfig | 10129 |
| -rw-r--r-- | Kconfig.binfmt | 7550 |
| -rw-r--r-- | Makefile | 4601 |
| d--------- | adfs | 423 |
| d--------- | affs | 461 |
| d--------- | afs | 1624 |
| -rw-r--r-- | aio.c | 62458 |
| -rw-r--r-- | anon_inodes.c | 8257 |
| -rw-r--r-- | attr.c | 13874 |
| d--------- | autofs | 359 |
| -rw-r--r-- | bad_inode.c | 5904 |
| d--------- | befs | 648 |
| d--------- | bfs | 206 |
| -rw-r--r-- | binfmt_aout.c | 8395 |
| -rw-r--r-- | binfmt_elf.c | 63050 |
| -rw-r--r-- | binfmt_elf_fdpic.c | 44465 |
| -rw-r--r-- | binfmt_elf_test.c | 2796 |
| -rw-r--r-- | binfmt_flat.c | 26377 |
| -rw-r--r-- | binfmt_misc.c | 17791 |
| -rw-r--r-- | binfmt_script.c | 4291 |

Figure 1.5 – The source for kernel 5.19.9

You'll notice source files such as `open.c` and `read_write.c`, which are the functions invoked when a user space process generates `open ()`, `read ()`, and `write ()` system calls. These files contain a lot of code, and since we won't create any new code here, this is merely a poking exercise. Nevertheless, there are a few important pieces of code in these files that highlight what we explained earlier. Let's take a quick peek at the read and write functions.

The `SYSCALL_DEFINE3` macro is the standard way to define a system call and takes the name of the system call as one of the parameters.

For the `write` system call, this definition looks as follows. Note that one of the parameters is the file descriptor:

```
SYSCALL_DEFINE3(write, unsigned int, fd, const char __user *, buf,
size_t, count)
{
        return ksys_write(fd, buf, count);
}
```

Similarly, this is the definition for the `read` system call:

```
SYSCALL_DEFINE3(read, unsigned int, fd, char __user *, buf, size_t,
count)
{
        return ksys_read(fd, buf, count);
}
```

Both call the `ksys_write ()` and `ksys_read ()` functions. Let's see the code for these two functions:

```
ssize_t ksys_read(unsigned int fd, char __user *buf, size_t count)
{
        struct fd f = fdget_pos(fd);
        ssize_t ret = -EBADF;
****** Skipped ******
                ret = vfs_read(f.file, buf, count, ppos);
****** Skipped ******
        return ret;
}
ssize_t ksys_write(unsigned int fd, const char __user *buf, size_t
count)
{
        struct fd f = fdget_pos(fd);
        ssize_t ret = -EBADF;
****** Skipped ******
```

```
                              ret = vfs_write(f.file, buf, count, ppos);
                              ****** Skipped ******
               return ret;
     }
```

The presence of the vfs_read () and vfs_write () functions indicates that we're transitioning to the VFS. These functions look up the file_operations structure for the underlying filesystem and invoke the appropriate read () and write () methods:

```
ssize_t vfs_read(struct file *file, char __user *buf, size_t count,
loff_t *pos)
{
****** Skipped ******
         if (file->f_op->read)
                 ret = file->f_op->read(file, buf, count, pos);
         else if (file->f_op->read_iter)
                 ret = new_sync_read(file, buf, count, pos);
****** Skipped ******
return ret;
}
ssize_t vfs_write(struct file *file, const char __user *buf, size_t
count, loff_t *pos)
{
****** Skipped ******
         if (file->f_op->write)
                 ret = file->f_op->write(file, buf, count, pos);
         else if (file->f_op->write_iter)
                 ret = new_sync_write(file, buf, count, pos);
   ****** Skipped ******
         return ret;
}
```

Each filesystem defines the file_operations structure of pointers for supporting operations. There are multiple definitions of the file_operations structure in the kernel source code, unique to each filesystem. The operations defined in this structure describe how read or write functions will be performed:

```
root@linuxbox:/linux-5.19.9/fs# grep -R "struct file_operations" * |
wc -l
453
root@linuxbox:/linux-5.19.9/fs# grep -R "struct file_operations" *
9p/vfs_dir.c:const struct file_operations v9fs_dir_operations = {
9p/vfs_dir.c:const struct file_operations v9fs_dir_operations_dotl = {
9p/v9fs_vfs.h:extern const struct file_operations v9fs_file_
operations;
```

```
9p/v9fs_vfs.h:extern const struct file_operations v9fs_file_
operations_dotl;
9p/v9fs_vfs.h:extern const struct file_operations v9fs_dir_operations;
9p/v9fs_vfs.h:extern const struct file_operations v9fs_dir_operations_
dotl;
9p/v9fs_vfs.h:extern const struct file_operations v9fs_cached_file_
operations;
9p/v9fs_vfs.h:extern const struct file_operations v9fs_cached_file_
operations_dotl;
9p/v9fs_vfs.h:extern const struct file_operations v9fs_mmap_file_
operations;
9p/v9fs_vfs.h:extern const struct file_operations v9fs_mmap_file_
operations_dotl;
```

[The rest of the code is skipped for brevity.]

As you can see, the `file_operations` structure is used for a wide range of file types, including regular files, directories, device files, and network sockets. In general, any type of file that can be opened and manipulated using standard file I/O operations can be covered by this structure.

## Tracing VFS functions

There are quite a few tracing mechanisms available in Linux that can offer a glance at how things work under the hood. One of them is the **BPF Compiler Collection (BCC)** tools. These tools offer a wide range of scripts that can record events for different subsystems in the kernel. You can install these tools for your operating system by following the instructions in the *Technical requirements* section. For now, we're just going to use one of the programs from this toolkit, called `funccount`. As the name suggests, `funccount` counts the number of function calls:

```
root@linuxbox:~# funccount --help
usage: funccount [-h] [-p PID] [-i INTERVAL] [-d DURATION] [-T] [-r]
[-D]
                 [-c CPU]
                 pattern
Count functions, tracepoints, and USDT probes
```

Just to test and verify our understanding of what we stated earlier, we're going to run a simple copy process in the background and use the `funccount` program to trace the VFS functions that are invoked as a result of the `cp` command. As we're going to count the VFS calls for the `cp` process only, we need to use the `-p` flag to specify a process ID. The `vfs_*` parameter will trace all the VFS functions for the process. You'll see that the `vfs_read ()` and `vfs_write ()` functions are invoked by the `cp` process. The COUNT column specifies the number of times the function was called:

```
funccount -p process_ID 'vfs_*'
[root@linuxbox ~]# nohup cp myfile /tmp/myfile &
```

```
[1] 1228433
[root@linuxbox ~]# nohup: ignoring input and appending output to
'nohup.out'
[root@linuxbox ~]#
[root@linuxbox ~]# funccount -p 1228433 "vfs_*"
Tracing 66 functions for "b'vfs_*'"... Hit Ctrl-C to end.
^C
FUNC                                      COUNT
b'vfs_read'                               28015
b'vfs_write'                              28510
Detaching...
[root@linuxbox ~]#
```

Let's run this again and see what system calls are used when doing a simple copy operation. As expected, the most frequently used system calls when doing cp are read and write:

```
funccount 't:syscalls:sys_enter_*' -p process_ID
[root@linuxbox ~]# nohup cp myfile /tmp/myfile &
[1] 1228433
[root@linuxbox ~]# nohup: ignoring input and appending output to
'nohup.out'

[root@linuxbox ~]#
[root@linuxbox ~]# /usr/share/bcc/tools/funccount -p 1228433 "vfs_*"
Tracing 66 functions for "b'vfs_*'"... Hit Ctrl-C to end.
^C
FUNC                                      COUNT
b'vfs_read'                               28015
b'vfs_write'                              28510
Detaching...
[root@linuxbox ~]#
```

Let's summarize what we covered in this section. Linux offers support for a wide range of filesystems, and the VFS layer in the kernel ensures that this can be achieved without any hassle. The VFS provides a standardized way for end user processes to interact with the different filesystems. This standardization is achieved by implementing a common file mode. The VFS defines several virtual functions for common file operations. As a result of this approach, applications can universally perform regular file operations. When a process generates a system call, the VFS will redirect these calls to the appropriate function of the filesystem.

# Explaining the *Everything is a file* philosophy

In Linux, all of the following are considered files:

- Directories

- Disk drives and their partitions

- Sockets

- Pipes

- CD-ROM

The phrase *everything is a file* implies that all the preceding entities in Linux are represented by file descriptors, abstracted over the VFS. You could also say that *everything has a file descriptor*, but let's not indulge in that debate.

The *everything is a file* ideology that characterizes the architecture of a Linux system is also implemented courtesy of the VFS. Earlier, we defined pseudo filesystems as filesystems that generate their content on the fly. These filesystems are also referred to as VFSes and play a major role in implementing this concept.

You can retrieve the list of filesystems currently registered with the kernel through the `procfs` pseudo filesystem. When seeing this list, note `nodev` in the first column against some filesystems. `nodev` indicates that this is a pseudo filesystem and is not backed by a block device. Filesystems such as Ext2, 3, and 4 are created on a block device; hence, they do not have the `nodev` entry in the first column:

```
cat /proc/filesystems
[root@linuxbox ~]# cat /proc/filesystems
nodev    sysfs
nodev    tmpfs
nodev    bdev
nodev    proc
nodev    cgroup
nodev    cgroup2
nodev    cpuset
nodev    devtmpfs
nodev    configfs
nodev    debugfs
nodev    tracefs
nodev    securityfs
nodev    sockfs
nodev    bpf
nodev    pipefs
nodev    ramfs
```

[The rest of the code is skipped for brevity.]

You can also use the mount command to find out about the currently mounted pseudo filesystems in your system:

```
mount | grep -v sd | grep -ivE ":/|mapper"
[root@linuxbox ~]# mount | grep -v sd | grep -ivE ":/|mapper"
sysfs on /sys type sysfs (rw,nosuid,nodev,noexec,relatime)
proc on /proc type proc (rw,nosuid,nodev,noexec,relatime)
devtmpfs on /dev type devtmpfs (rw,nosuid,size=1993552k,nr_
inodes=498388,mode=755)
securityfs on /sys/kernel/security type securityfs (rw,nosuid,nodev,-
noexec,relatime)
tmpfs on /dev/shm type tmpfs (rw,nosuid,nodev)
devpts on /dev/pts type devpts (rw,nosuid,noexec,relatime,gid=5,-
mode=620,ptmxmode=000)
tmpfs on /run type tmpfs (rw,nosuid,nodev,mode=755)
tmpfs on /sys/fs/cgroup type tmpfs (ro,nosuid,nodev,noexec,mode=755)
cgroup on /sys/fs/cgroup/systemd type cgroup (rw,nosuid,nodev,noex-
ec,relatime,xattr,release_agent=/usr/lib/systemd/systemd-cgroups-
agent,name=systemd)
pstore on /sys/fs/pstore type pstore (rw,nosuid,nodev,noexec,relatime)
efivarfs on /sys/firmware/efi/efivars type efivarfs (rw,nosuid,nodev,-
noexec,relatime)
```

[The rest of the code is skipped for brevity.]

Let's take a tour of the /proc directory. You'll see a long list of numbered directories; these numbers represent the IDs of all the processes that currently run on your system:

```
[root@linuxbox ~]# ls /proc/
1     1116   1228072  1235  1534  196  216  30   54   6   631  668  81
0     ioports        scsi
10    1121   1228220  1243  1535  197  217  32   55   600  632  670
9     irq            self
1038  1125   1228371  1264  1536  198  218  345  56   602
633   673   905      kallsyms      slabinfo
1039  1127   1228376  13    1537  199  219  347  570  603
634   675   91       kcore         softirqs
1040  1197   1228378  14    1538  2    22   348  573  605  635  677  94
7     keys           stat
1041  12     1228379  1442  16    20   220  37   574  607
636   679   acpi     key-users  swaps
1042  1205   1228385  1443  1604  200  221  38   576  609
637   681   buddyinfo  kmsg        sys
1043  1213   1228386  1444  1611  201  222  39   577  610  638  684  bu
s     kpagecgroup  sysrq-
```

[The rest of the code is skipped for brevity.]

The `procfs` filesystem offers us a glimpse into the running state of the kernel. The content in `/proc` is generated when we want to view this information. This information is not persistently present on your disk drive. This all happens in memory. As you can see from the `ls` command, the size of `/proc` on disk is zero bytes:

```
[root@linuxbox ~]# ls -ld /proc/
dr-xr-xr-x 292 root root 0 Sep 20 00:41 /proc/
[root@linuxbox ~]#
```

`/proc` provides an on-the-spot view of the processes running on the system. Consider the `/proc/cpuinfo` file. This file displays the processor-related information for your system. If we check this file, it will be shown as `empty`:

```
[root@linuxbox ~]# ls -l /proc/cpuinfo
-r--r--r-- 1 root root 0 Nov  5 02:02 /proc/cpuinfo
[root@linuxbox ~]#
[root@linuxbox ~]# file /proc/cpuinfo
/proc/cpuinfo: empty
[root@linuxbox ~]#
```

However, when the file contents are viewed through `cat`, they show a lot of information:

```
[root@linuxbox ~]# cat /proc/cpuinfo
processor       : 0
vendor_id       : GenuineIntel
cpu family      : 6
model           : 79
model name      : Intel(R) Xeon(R) CPU E5-2683 v4 @ 2.10GHz
stepping        : 1
microcode       : 0xb00003e
cpu MHz         : 2099.998
cache size      : 40960 KB
physical id     : 0
siblings        : 1
core id         : 0
cpu cores       : 1
apicid          : 0
initial apicid  : 0
fpu             : yes
fpu_exception   : yes
cpuid level     : 20
wp              : yes
```

[The rest of the code is skipped for brevity.]

Linux abstracts all entities such as processes, directories, network sockets, and storage devices into the VFS. Through the VFS, we can retrieve information from the kernel. Most Linux distributions offer a variety of tools for monitoring the consumption of storage, compute, and memory resources. All these tools gather stats for various metrics through the data available in `procfs`. For instance, the `mpstat` command, which provides stats about all the processors in a system, retrieves data from the `/proc/stat` file. It then presents this data in a human-readable format for a better understanding:

```
[root@linuxbox ~]# cat /proc/stat
cpu  5441359 345061 4902126 1576734730 46546 1375926 942018 0 0 0
cpu0 1276258 81287 1176897 394542528 13159 255659 280236 0 0 0
cpu1 1455759 126524 1299970 394192241 13392 314865 178446 0 0 0
cpu2 1445048 126489 1319450 394145153 12496 318550 186289 0 0 0
cpu3 1264293 10760 1105807 393854806 7498 486850 297045 0 0 0
```

[The rest of the code is skipped for brevity.]

If we use the `strace` utility on the `mpstat` command, it will show that under the hood, `mpstat` uses the `/proc/stat` file to display processor stats:

```
strace mpstat 2>&1 |grep "/proc/stat"
[root@linuxbox ~]# strace mpstat 2>&1 |grep "/proc/stat"
openat(AT_FDCWD, "/proc/stat", O_RDONLY) = 3
[root@linuxbox ~]#
```

Similarly, popular commands such as `top`, `ps`, and `free` gather memory-related information from the `/proc/meminfo` file:

```
[root@linuxbox ~]# strace free -h 2>&1 |grep meminfo
openat(AT_FDCWD, "/proc/meminfo", O_RDONLY) = 3
[root@linuxbox ~]#
```

Similar to `/proc`, another commonly used pseudo filesystem is **sysfs**, which is mounted at `/sys`. The **sysfs filesystem** mostly contains information about hardware devices on your system. For example, to find information about the disk drive in your system, such as its model, you can issue the following command:

```
cat /sys/block/sda/device/model
[root@linuxbox ~]# cat /sys/block/sda/device/model
SAMSUNG MZMTE512
[root@linuxbox ~]#
```

Even LEDs on a keyboard have a corresponding file in `/sys`:

```
[root@linuxbox ~]# ls /sys/class/leds
ath9k-phy0  input4::capslock  input4::numlock  input4::scrolllock
[root@linuxbox ~]#
```

The *everything is a file* philosophy is one of the defining features of the Linux kernel. It signifies that everything in a system, including regular text files, directories, and devices, can be abstracted over the VFS layer in the kernel. As a result, all these entities can be represented as file-like objects through the VFS layer. There are several pseudo filesystems in Linux that contain information about the different kernel subsystems. The content of these pseudo filesystems is only present in memory and generated dynamically.

## Summary

The Linux storage stack is a complex design and consists of multiple layers, all of which work in coordination. Like other hardware resources, storage lies in the kernel space. When a user space program wants to access any of these resources, it has to invoke a system call. The system call interface in Linux allows user space programs to access resources in the kernel space. When a user space program wants to access something on the disk, the first component it interacts with is the VFS subsystem. The VFS provides an abstraction of filesystem-related interfaces and is responsible for accommodating multiple filesystems in the kernel. Through its common filesystem interface, the VFS intercepts the generic system calls (such as `read()` and `write()`) from the user space programs and redirects them to their appropriate interfaces in the filesystem layer. Because of this approach, the user space programs do not need to worry about the underlying filesystems being used, and they can uniformly perform filesystem operations.

This chapter served as an introduction to the major Linux kernel subsystem Virtual Filesystem and its primary functions in the Linux kernel. The VFS provides a common interface for all filesystems through data structures, such as inodes, superblocks, and directory entries. In the next chapter, we will take a look at these data structures and explain how they all help the VFS to manage multiple filesystems.

# 2

# Explaining the Data Structures in a VFS

In the first chapter of this book, we got a good look at the **virtual filesystem** (**VFS**), its most common functions, why it is necessary, and how it plays a pivotal role in implementing the *everything is a file* concept in Linux. We also explained the system call interface in Linux and how user-space applications can use generic system calls and interact with the VFS. The VFS is sandwiched between user-space programs and actual filesystems and implements a common file model so that applications can use uniform access methods to perform their operations, regardless of the filesystems being used.

While talking about the different filesystems, we mentioned that the VFS uses structures such as inodes, superblocks, and directory entries to represent a generic view of the filesystems. These structures are crucial as they ensure a clear distinction between the metadata and the actual data of a file.

This chapter will introduce you to the different data structures in the kernel's VFS. You will get to know how the kernel uses structures such as inodes and directory entries to store metadata about files. You will also learn how the kernel is able to record the filesystem characteristics through the superblock structure. At the end, we'll explain the caching mechanisms in VFS.

We're going to cover the following main topics:

- Inodes
- Superblocks
- Directory entries
- File objects
- Page cache

## Technical requirements

It would be helpful to have a decent understanding of Linux operating system concepts. This includes knowledge of filesystems, processes, and memory management. We're not going to create any new code in this book but if you want to explore the Linux kernel in more detail, understanding C programming concepts is crucial for comprehending VFS data structures. As a general rule, you should make it a habit to refer to the official kernel documentation as it can provide in-depth information about the kernel's internal workings.

The commands and examples presented in this chapter are distribution agnostic and can be run on any Linux operating system, such as Debian, Ubuntu, Red Hat, Fedora, and so on. There are a few references to the kernel source code. If you want to download the kernel source, you can download it from `https://www.kernel.org`. The code segments referred to in this chapter and book are from kernel `5.19.9`.

## Data structures in VFS

VFS uses several data structures to implement generic abstraction methods for all filesystems and provides the filesystem interface to user-space programs. These structures ensure a certain amount of commonality between the design and operations of filesystems. One important point to remember is that all the methods defined by VFS are not enforced upon filesystems. Yes, the filesystems should adhere to the structures defined in VFS and build upon them to ensure commonality among them. But there might be a lot of methods and fields in these structures that are not applicable to a particular filesystem. In such cases, filesystems stick to the relevant fields as per their design and leave out the surplus information. As we're going to explain common VFS data structures, it is imperative that we look at the relevant code segments in the kernel for some clarification. Nevertheless, I've tried my best to present the material in a generic way so that most concepts can be understood even without developing an understanding of the code.

Ancient Greeks believed that four elements made up everything: earth, water, air, and fire. Likewise, the following structures make up VFS – well, most of it:

- Inodes
- Directory entries
- File objects
- Superblocks

# Inodes – indexing files and directories

When storing data on disk, Linux follows one strict rule: *all the outside-of-the-envelope information must be kept apart from the contents inside the envelope*. In other words, the data describing a file is isolated from the actual data in the file. The structure that holds this metadata is called the **index node**, shortened as **inode**. The inode structure contains metadata for files and directories in Linux. The name of a file or directory is merely a pointer to an inode, and each file or directory has exactly one inode.

Consider the Marauder's Map as an analogy (*Harry Potter*, anyone?). The map shows the location of every person in the school. Each person is represented by a dot on the map, and when you click on the dot, it reveals information about the person, such as their name, location, and status. Think of the Marauder's Map as the filesystem, and the dots representing people as the inodes showing metadata.

But what constitutes the metadata for a file? When you do a simple listing of a file through the ls command, you see a bunch of information, such as file permissions, ownership, time stamps, and so on. All these details constitute the metadata of the file since they are describing some properties of the file, not its actual contents.

Some of the file metadata can be checked through a simple ls command. Although a slightly better command for displaying the metadata of a file is stat, as it provides a lot more information about the file attributes. For instance, it shows the access, modification, change timestamps, the device where the file is located, the number of blocks reserved on the drive for the file, and the inode number of the file.

If want to get detailed information about a file's metadata, such as /etc/hosts, we can use the stat command as follows:

```
stat /etc/hosts
```

Note the inode number (67118958) of /etc/hosts in the output of the stat command:

```
[root@linuxbox ~]# stat /etc/hosts
  File: /etc/hosts
  Size: 220             Blocks: 8           IO Block: 4096    regular
file
Device: fd00h/64768d    Inode: 67118958     Links: 1
Access: (0644/-rw-r--r--)  Uid: (    0/    root)   Gid:
(    0/    root)
Access: 2022-11-20 04:00:38.054988422 -0500
Modify: 2022-06-15 22:30:32.755324938 -0400
Change: 2022-06-15 22:30:32.755324938 -0400
Birth: 2022-06-15 22:30:32.755324938 -0400
[root@linuxbox ~]#
```

An inode number works as a unique identifier for a file. Consider the example of the **Domain Name System (DNS)**. We use human-readable website names so that we don't have to remember the IP address of each website. Similarly, file and directory names ensure that we don't need to remember the inode number. The kernel keeps track of each file and directory through its inode. An inode can be considered a low-level name for a file. For instance, it is possible to locate a file through its inode number. The find command provides the inum argument to search for a file through its inode:

```
find / -inum 67118958 -exec ls -l {} \;
```

If we search for the inode number we retrieved from the stat command, we can retrieve the corresponding file:

```
[root@linuxbox ~]# find / -inum 67118958 -exec ls -l {} \;
-rw-r--r-- 1 root root 220 Jun 15 22:30 /etc/hosts
[root@linuxbox ~]#
```

Inodes are only unique within a filesystem boundary. If directories on your system (such as /home and /tmp) are on separate disk partitions and filesystems, the same inode number could be assigned to a different file on each filesystem:

```
[root@linuxbox ~]# ls -li /home/pokemon/pikachu
134460858 -rw-r--r-- 1 root root 1472 Oct 11 05:10 /home/pokemon/
pikachu
[root@linuxbox ~]#
[root@linuxbox ~]# ls -li /tmp/bulbasaur
134460858 -rw-r--r-- 1 root root 259 Nov 20 04:36 /tmp/bulbasaur
[root@linuxbox ~]#
```

The uniqueness of inode numbers within filesystem boundaries ties to the concept of **linking**. As the same inode number can be used by different filesystems, hard links do not cross filesystem boundaries.

## Defining an inode in the kernel

In the kernel source code, the definition of an inode is present in linux/fs.h. There are an innumerable number of fields in this definition. Please note that this definition of **struct inode** is general and all-encompassing. An inode is a filesystem-specific property. It is not obligatory for a filesystem to define all these fields in its inode definition. The definition for the inode structure is fairly long and, as such, we're going to limit ourselves to some basic fields:

```
struct inode {
        umode_t              i_mode;
        unsigned short       i_opflags;
        kuid_t               i_uid;
        kgid_t               i_gid;
        unsigned int         i_flags;
```

```
        const struct inode_operations    *i_op;
        struct super_block       *i_sb;
[........................ . . .]
```

Some fields of interest are defined here:

- i_mapping: This is a pointer to the address space structure that holds the mappings for the inode's data blocks. This field is initialized by the filesystem when an inode is created or when it is read from disk. For instance, when a process writes data to a file, the kernel uses this field to map the appropriate memory pages to the file's data blocks. (The data blocks are explained in the next sections.)

- i_uid and i_gid: These are for the user and group owner, respectively.

- i_flags: This defines filesystem-specific flags.

- i_acl: This is for access control lists for filesystems.

- i_op: This points to the inode operations structure, which defines all the operations that can be performed on an inode, such as creating, reading, writing, and modifying the file attributes.

- i_sb: This is pointing to the superblock structure of the underlying filesystem where the inode resides. (There's a separate topic for explaining the superblock structure.)

- i_rdev: This field stores the device number for some special files. For instance, the kernel creates special files to represent hard disks and other devices in the system. When a special file is created, the kernel assigns it a unique device number, creates an inode for the device, and sets this field to point to the device's identifier.

- i_atime, i_mtime, and i_ctime: These are the access, modified, and change timestamps, respectively.

- i_bytes: This is the number of bytes in the file.

- i_blkbits: This field stores the number of bits needed to represent the block size of the filesystem to which the inode belongs.

- i_blocks: This field stores the total number of disk blocks used by the file represented by the inode.

- i_fop: This is a pointer to the file operations structure associated with the inode. For instance, when a process opens a file, the kernel uses this field to obtain a pointer to the file operations structure for that file. It can then use the functions defined in the file operations structure to perform operations on the file, such as reading or writing.

- i_count: This is used to keep track of the number of active references to the inode. Whenever a new process accesses a file, this counter is incremented for that file. If this field reaches a value of zero, it means that there are no more references to the inode, and it can be safely deallocated.

- `i_nlink`: This field references the number of hard links to the inode.
- `i_io_list`: This is a list used to track inodes that have pending I/O requests. When the kernel adds an I/O request to the queue for an inode, that inode is added to this list. When the I/O request has been completed, the inode is removed from this list.

There are around 50 fields in the definition for the inode structure, so we've barely scratched the surface here. But this should give us an idea that the inode defines much more than surface-level information for a file. Don't worry if you're confused. We're going to explain inodes in a lot more detail. There are two types of operations applicable to an inode structure, which are defined by `file_operations` and `inode_operations` structures. We'll go through the `file_operation` structure a bit later when we cover file objects in the *File objects – representing open files* section.

## Defining inode operations

The inode operations structure contains a set of function pointers that define how the filesystem interacts with inodes. Each filesystem has its own inode operations structure, which is registered with the VFS when the filesystem is mounted.

The `inode_operations` struct is referred to by the `i_op` pointer. Remember when we explained the *everything is a file* concept in *Chapter 1, Where It All Starts From – The Virtual Filesystem* Since everything is a file, albeit of a different type, an inode is assigned to each of them. Disk drives, disk partitions, regular text files, documents, pipes, and sockets all have an inode assigned to them. There's an inode for every directory as well. But all these *files* are of a different nature and represent different entities in your system. For instance, the inode operations applicable to a directory are different than a regular text file. The `inode_operations` structure provides all the functions that an inode needs to implement for each type of file, for managing inode data.

Each inode is associated with an instance of the `inode_operations` structure, which provides a set of operations that can be performed on the inode. This structure contains pointers to different functions that are used to manipulate inodes:

```
struct inode_operations {
        struct dentry * (*lookup) (struct inode *,struct dentry *,
unsigned int);
        const char * (*get_link) (struct dentry *, struct inode *,
struct delayed_call *);
        int (*permission) (struct user_namespace *, struct inode *,
int);
        struct posix_acl * (*get_acl)(struct inode *, int, bool);
        int (*readlink) (struct dentry *, char __user *,int);
        int (*create) (struct user_namespace *, struct inode *,struct
dentry *, umode_t, bool);
        int (*link) (struct dentry *,struct inode *,struct dentry *);
        int (*unlink) (struct inode *,struct dentry *);
```

```
            int (*symlink) (struct user_namespace *, struct inode *,struct
dentry *,
    [.......................................]
```

Some of the important operations that can be performed using this structure are described here:

- `lookup`: This is used for searching an inode entry in a directory. It takes a directory inode and a filename as arguments, and it returns a pointer to the inode that corresponds to the filename.

- `create`: This function is called when a new file or directory is created, and it is responsible for initializing the inode with the appropriate metadata, such as ownership and permissions. This is used for constructing an inode object in response to an `open ()` system call.

- `get_link`: This is used for working with symbolic links. A symbolic link is pointing to another inode.

- `permission`: When a file is to be accessed, VFS invokes this function to check for access rights on the file.

- `link`: This is invoked in response to the `link ()` system call, which creates a new hard link. It increments the link count of the inode and updates its metadata.

- `symlink`: This is invoked in response to the `symlink ()` system call, which creates a new soft link.

- `unlink`: This is invoked in response to the `unlink ()` system call and deletes the file link. It decrements the link count of the inode and deletes the inode from the disk if the link count reaches zero.

- `mkdir` and `rmdir`: These are invoked in response to `mkdir ()` and `rmdir ()` system calls for creating and deleting directories, respectively.

## Tracking file data on disk through inodes

Since every file in the system is going to have some metadata, it will always have exactly one inode associated with it. As every inode is storing some information, filesystems need to reserve some space for them, typically just a few bytes. For instance, the Ext4 filesystem by default uses 256 bytes for a single inode. Filesystems maintain an **inode table** to keep track of used and free inodes.

The fields present in an inode structure provide the following two types of information about a file:

- **File attributes**: Details about file ownership, permissions, timestamps, links, and the number of blocks used

- **Data blocks**: Pointers to data blocks on disk, where the actual file content is stored

In addition to file permissions, ownership, and timestamps, another important piece of information provided by the inode is the location of actual data on the disk. A file can span across multiple disk blocks, depending on its size. The inode structure uses pointers to track this information. Why is this necessary? This tracking of disk blocks is required as there is no guarantee that the data in a file will be stored and accessed in a sequential or contiguous manner. The pointers used by an inode are typically 4 bytes in size and can be classified as direct and indirect pointers. For smaller files, an inode contains direct pointers to the data blocks of a file. Each direct pointer points to the disk address that is storing file data.

Using direct pointers for referring to disk addresses was always going to have a major limitation. The question was: how many direct pointers are enough? File sizes can vary from a few bytes to terabytes. Using 15 direct pointers in the structure meant that for a block size of 4 KB, we could only point to 60 KB of data. Of course, this wouldn't work in any dimension, as even small text files tend to be larger than 60 KB. This is depicted in *Figure 2.1*:

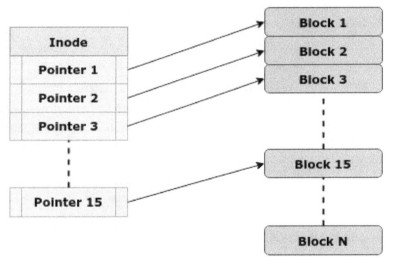

Figure 2.1 – Limitation when using direct pointers: for a block
size of 4 KB, only 60 KB of data can be addressed

To cope with this problem, indirect pointers are used. An inode structure contains 12 direct and 3 indirect pointers. Unlike a direct pointer, an **indirect pointer** is a pointer to a block of pointers. When all direct pointers have been exhausted, the filesystem uses a data block to store additional pointers. The unlucky 13th or single indirect pointer in an inode points to this data block. The pointers inside this block point to the data blocks, which actually contain the file data. When the file size cannot be addressed using the single indirect pointer, double indirect pointers are used. **Double indirect pointers** point to a block that contains pointers to indirect blocks, each of which contains pointers to on-disk addresses. Similarly, when the file grows beyond the limits of a double indirect pointer—yes, you guessed it—**triple indirect pointers** are used.

At this point, you're probably off your head, and you're thinking there is no point(-er). Needless to say, this entire hierarchy is pretty complex. Some modern filesystems make use of a concept called **extents** for storing even larger files. We're going to cover that when we cover block filesystems in *Chapter 3, Exploring the Actual Filesystems Under the VFS*.

For now, let us simplify this and point ourselves in the right direction. We're going to make use of some basic math to explain how indirect pointers help in storing larger files. We're going to consider a block size of 4 KB, as this is the default used by most filesystems:

- Total pointers in an inode = 15
- Number of direct pointers = 12
- Number of indirect pointers = 1
- Number of double indirect pointers = 1
- Number of triple indirect pointers = 1
- Size of each pointer (direct or indirect) = 4 bytes
- Number of pointers per block = (block size) / (pointer size) = (4 KB / 4) = 1,024 pointers
- Maximum file size that can be referred by using direct pointers = 12 x 4 KB = 48 KB
- Maximum file size that can be referred by using 12 direct and 1 indirect pointer = [(12 x 4 KB) + (1,024 x 4 KB)] ≈ 4 MB
- Maximum file size that can be referred by using 12 direct, 1 indirect, and one double indirect pointer = [(12 x 4 KB) + (1,024 x 4 KB) + (1,024 x 1,024 x 4 KB)] ≈ 4 GB
- Maximum file size that can be referred by using 12 direct pointers, 1 single indirect, 1 double indirect, and 1 triple indirect pointer = (12 x 4 KB) + (1,024 x 4 KB) + (1,024 x 1,024 x 4 KB) + (1 x 1,024 x 1,024 x 1,024 x 4 KB) ≈ 4 TB

The following figure shows how the use of indirect pointers can help in addressing larger files:

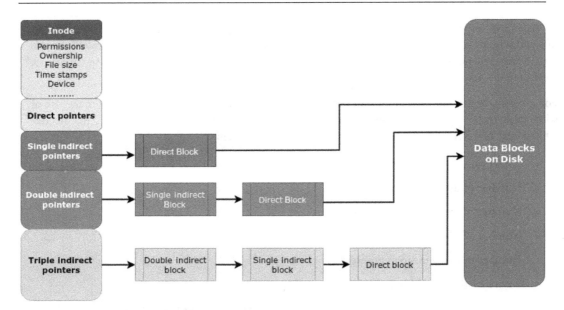

Figure 2.2 – Visual representation of an inode structure

As shown in *Figure 2.2*, filesystems might use a single-level indirect block for smaller files, and then switch to a double-level indirect block for larger files. Using indirect inode pointers offers multiple advantages. First, it eliminates the need for contiguous storage allocation to accommodate large files, thereby enabling the filesystem to handle such files effectively. Second, it facilitates efficient space utilization since blocks can be allocated for a file on an as-needed basis rather than the upfront reservation of a significant amount of space. Each inode typically has 12 direct block pointers, 1 single indirect block pointer, 1 double indirect block pointer, and 1 triple indirect block pointer.

## Can a filesystem run out of inodes?

When managing storage, keeping space available is a major concern. Running out of disk space is a common scenario. An inode is assigned to every file and directory on the filesystem, but what if all inode numbers have been assigned? It's highly unlikely, as filesystems usually have several millions of inodes available. But yes, it is possible for a filesystem to run out of inodes. And if it does, the amount of free space on the disk won't be of any use as the filesystem won't be able to create any new file. The number of inodes in a filesystem cannot be expanded once the filesystem has been created, so a backup will be the only option. You can check the inode usage of mounted filesystems using the df command:

```
df -Thi
```

This situation is illustrated in *Figure 2.3*. The filesystem mounted on /ford has close to 40% of free space, but since all the 6.3 million inodes have been exhausted, it is not possible to create any new file:

```
[root@linuxbox ~]# touch /ford/driver
touch: cannot touch '/ford/driver': No space left on device
[root@linuxbox ~]#
[root@linuxbox ~]# df -Th /ford
Filesystem      Type   Size   Used Avail Use% Mounted on
/dev/sdc        ext4   13G    7.7G  4.8G  62%  /ford
[root@linuxbox ~]#
[root@linuxbox ~]# df -Thi /ford
Filesystem      Type   Inodes IUsed IFree IUse% Mounted on
/dev/sdc        ext4   6.3M   6.3M     0  100%     /ford
[root@linuxbox ~]#
```

Figure 2.3 – A filesystem can run out of inodes

Let us wrap up our discussion with a few key inode pointers:

- In addition to metadata, inodes also save information about where a file is stored on the physical disk. In order to keep track of a file's physical whereabouts, inodes use several direct and indirect pointers.

- Inodes are stored in on-disk filesystem structures in an inode table. When a file needs to be opened, its corresponding inode is loaded into memory.

- For filesystems that only generate their content in memory, such as procfs and sysfs, their inodes are only present in memory.

- While inodes store a lot of metadata about a file, they do not store the name of the file. So, two things that are not part of the inode structure are the file contents and the filename.

- Inodes use 15 pointers to keep track of a file's data blocks on a disk. The first 12 pointers are direct pointers and can only address a maximum file size of 48 KB. The remaining three pointers provide a single, double, and triple level of indirection. Through the use of these indirect pointers, large files can be addressed.

- If a filesystem runs out of inodes, no new file can be created on it. This is very rare, as a filesystem usually has a large number of inodes, amounting to millions.

## Directory entries – mapping inodes to filenames

A directory acts as a catalog or a container for user files. The operations that are applicable to a directory are different than regular files. There are different commands for working with directories. A file is always going to be *inside* a directory, and to access that file, you need to specify the absolute or relative path in terms of directories. But like most things in Linux, directories are also treated as files. So, how does this all work?

Native Linux filesystems treat directories as files and store them like files. Like all regular files, a directory is also assigned an inode. There is one difference between the inode of a directory and a file. In the case of a directory, the **type** field in an inode is a **directory**. Remember, from our discussion about inodes, that an inode contains a lot of metadata about a file, but it doesn't contain the name of the file. The filename is present in a directory. A directory can be thought of as a special file that contains a table. This table consists of filenames and their corresponding inode numbers.

When trying to access a file, a process has to traverse the hierarchical directory structure. Each level in that structure defines a path that can either be absolute or relative. **Absolute pathnames**, also called **fully qualified pathnames**, begin with the root directory, such as /etc/ssh/sshd_config. In contrast, **relative pathnames** start with the current working directory of the process or user, such as ssh/sshd_config. There is no leading / in the relative pathnames.

*Fully qualified, mapping of names to numbers*, notice how this sounds a bit familiar to the concept of name resolution. While describing inodes, we used the analogy of the DNS, and we'll use it here as well. Just as regular DNS records map website names to IP addresses, a directory maps all filenames to their corresponding inode numbers. This combination of filenames to inodes is known as **linking**. The concept of linking is explained toward the end of this chapter.

If you prefer a pop culture reference, think of the star maps in the *Star Wars* franchise. In order to travel to a specific planet, the characters consult the star map to find the right location. Think of the star map as a directory, and each planet as a file or a subdirectory. The map lists the exact location and coordinates of each planet, much like a directory that lists the inode number and location of each file.

This mapping comes in handy when performing lookup operations. Looking up pathnames is a directory-centric operation as files are always present inside a directory. For looking up pathnames, VFS uses directory entries, also known as dentry objects. The dentry objects are responsible for depicting a directory in memory. When traversing a path, each component is considered a dentry object. If we take the example of the /etc/hosts file, the /etc directory and hosts file are both considered dentry objects and are mapped in memory. This helps in caching results of lookup operations, which in turn speeds up the overall performance when looking up pathnames.

Consider the following example: there is a /cars directory in the / partition, which contains three files: McLaren, Porsche, and Lamborghini. The following steps provide an oversimplified version of events that take place when a process wants to access the McLaren file in the /cars directory:

- The VFS will remodel the path as a dentry object.

- A dentry object will be created for each component in the pathname. The VFS will follow each directory entry for path resolution. For looking up /cars/McLaren, separate dentry objects will be created for /, cars, and McLaren.

- As our process has specified the absolute path, the VFS will start with the first component in the pathname, that is, /, and then proceed to the child objects.

- The VFS will check the relevant permissions on the inode to see whether the calling process has the required privileges.

- The VFS also calculates a hash value for `dentry objects` and compares it to the values in the hash table.

- The `/` directory contains the mapping of files and subdirectories to their respective inode numbers. Once the inode of `/cars` has been retrieved, the kernel can use the block pointers to view the on-disk contents of the directory.

- The `/cars` directory will contain the mapping of the three files (`McLaren`, `Porsche`, and `Lamborghini`) to their inode numbers. From here, we can use the inode of `McLaren`, which will point us to the on-disk data blocks that contain the file data.

It's important to know that the representation of a directory through `dentry objects` only exists in memory. They're not stored on disk. These objects are created by VFS on the fly:

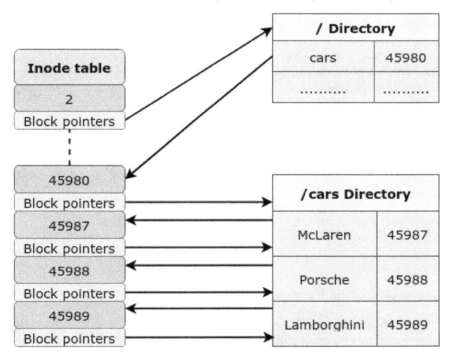

Figure 2.4 – The exchange between directories and inodes

The eagle-eyed reader might be wondering, how did we know the inode of the `/` directory? Most filesystems start assigning inode values from 2. The inode number 0 is not used. Inode number 1 is used to keep track of bad and defective blocks on the physical disk. So, inode allocation in filesystems starts from 2, and the root directory of a filesystem is always assigned inode number 2.

## Dentry cache

In terms of performance, pathname traversals and directory lookups can be expensive operations, especially if there are multiple recursive paths that need to be resolved. Once a path has been resolved, and a process wants to access the same path once again, the VFS will have to perform the entire operation again, which is excessive. There's also a dependency on the underlying storage media: how quickly it can retrieve the required information. This puts brakes on the operation.

We're once again going to use our go-to analogy here, the DNS. When a DNS client performs the same DNS query, the client's local DNS server caches the results of the query. This is done so that for any identical request, the DNS server wouldn't have to travel across the entire hierarchy of DNS servers. On similar lines, to speed up the lookup process of pathnames, the kernel uses the **directory entry cache**. Frequently accessed pathnames are kept in the memory to accelerate the lookup process. This saves a lot of unnecessary I/O requests to the underlying filesystem. The dentry cache plays a pivotal role in the filename lookup operation.

A directory maps filenames to inodes. You have to ask: if the dentry object is creating an in-memory representation of directories and results of lookup operations are being cached, that means corresponding inodes are also being cached? The answer is in the affirmative. There's no point in caching one without the other. If a directory entry is cached, the corresponding inode is also cached. The dentry objects pin corresponding inodes in memory, and they remain in memory as long as the dentry objects.

Dentry objects are defined in the `include/linux/dcache.h` file:

```
struct dentry {
        unsigned int d_flags;            /* protected by d_lock */
        seqcount_spinlock_t d_seq;       /* per dentry seqlock */
        struct hlist_bl_node d_hash;     /* lookup hash list */
        struct dentry *d_parent;         /* parent directory */
        struct qstr d_name;
        struct inode *d_inode;           /* Where the name belongs to -
NU
[..............]
```

Some commonly used terms are described here:

- d_name: This field contains a pointer to a `struct qstr` object, which represents the name of the file or directory. The `qstr` object is a structure used by the kernel to represent a string or sequence of characters.

- d_parent: This field contains a pointer to the parent directory of the file or directory associated with the directory entry.

- d_inode: This field is a pointer to the `struct inode` object of the file or directory.

- `d_lock`: This field contains a spinlock used to protect access to the `struct dentry` object. It is quite common that the `dentry` and `inode` objects are shared among multiple processes that open the same file or directory. The `d_lock` field protects these objects from concurrent modifications that could lead to inconsistent or corrupt filesystem data.

- `d_op`: This field contains a pointer to the `struct dentry_operations` structure, which contains a set of function pointers that define the operations that can be performed on the `dentry` object.

- `d_sb`: This is a pointer to the `struct super_block` structure, which defines the superblock of the filesystem that the directory entry belongs to.

The cache is represented in memory using a hash table. Each entry in the hash table structure points to a list of directory cache entries, having the same hash value. When a process attempts to access a file or a directory, the kernel searches the dentry cache for the corresponding directory entry, using the file or directory name as a key. If the entry is found in the cache, it is returned to the calling process. If the entry is not found, the kernel must go to disk and perform an I/O operation to read the directory entry from the filesystem.

## Dentry states

Dentry objects tend to be in one of the following three states:

- **Used**: A used dentry indicates a dentry object that is currently being used by the VFS and shows that there is a valid inode structure associated with it. This means that a process is actively using this entry.

- **Unused**: An unused entry also has a valid inode associated with it, but it is not being used by the VFS. If a pathname lookup operation (related to this entry) is performed again, that operation can be completed using this cached entry. If a need arises to reclaim memory, then this entry can be disposed of.

- **Negative**: A negative state is a bit unique in that it is a representation of a lookup operation that failed. For instance, if the file to be accessed has already been deleted or if the pathname doesn't exist to begin with, a typical **No such file or directory** message is returned to the calling process. As a result of this failed lookup, the VFS will create a negative dentry. Too many failed lookup operations can create unnecessary negative dentries and can adversely affect performance.

## Dentry operations

The various filesystem-related operations that can be performed on dentry objects are defined in the `dentry_operations` structure:

```
struct dentry_operations {
        int (*d_revalidate)(struct dentry *, unsigned int);
```

```
        int (*d_weak_revalidate)(struct dentry *, unsigned int);
        int (*d_hash)(const struct dentry *, struct qstr *);
        int (*d_compare)(const struct dentry *, unsigned int, const
char *, const struct qstr *);
        int (*d_delete)(const struct dentry *);
        int (*d_init)(struct dentry *);
        void (*d_release)(struct dentry *);
        void (*d_prune)(struct dentry *);
        [............................ .]
```

A few important operations are described as follows:

- d_revalidate: It can often happen that the dentry objects in the cache can become out of sync with the on-disk data. This is often true in the case of network filesystems. The kernel is dependent on the network to gather information about the on-disk structures. In such cases, VFS uses the d_revalidate operation to revalidate a dentry.

- d_weak_revalidate: When a path lookup operation ends at a dentry that was not obtained after a lookup in the parent directory, VFS calls the d_weak_revalidate operation.

- d_hash: This is used to calculate the hash value of a dentry. It takes a dentry as input and returns a hash value, which is used to look up the dentry in the directory cache.

- d_compare: This is used to compare the filenames of two dentries. It takes two dentries as arguments and returns true if they refer to the same file or directory, or false if they are different.

- d_init: This is called when initializing a dentry object.

- d_release: This is called when a dentry has to be deallocated. It frees the memory used by the dentry and any associated resources, such as cached data.

- d_iput: This is invoked when a dentry object loses its inode. This is called just before d_release.

- d_dname: This is used for generating pathnames for pseudo filesystems, such as procfs and sysfs.

Let us summarize our discussion about directory entries:

- Linux treats directories as files. Directories also have an inode assigned to them. The name of a file is stored in the directory.

- The difference between the inode of a file and a directory lies in the contents of their corresponding disk blocks. The on-disk data of a directory contains a mapping of filenames and their inode numbers.

- Directories are represented in memory through dentry objects. The dentry objects are created by VFS in memory and are not stored on the physical disk.

- To optimize lookup operations, a dentry cache is used. The dentry cache keeps the recently accessed pathnames and their inodes in memory.

# File objects – representing open files

Similar to dentry objects, **file objects** are an in-memory representation of open files. The file object represents a process's view of an open file. Like dentry objects, file objects also do not correspond to any on-disk structures. When we covered system calls in *Chapter 1*, it was mentioned that the user-space programs interact with VFS through the system call interface. These system calls are generic functions for performing common operations such as `read` and `write`. The idea behind all this is to ensure that user-space programs don't have to worry about filesystems and their data structures.

When applications generate a system call to access a file, such as `open ()`, a file object gets created in memory. Similarly, when the application no longer needs access to the file and decides to close it using `close ()`, the file object is discarded. It's important to note that VFS can create multiple file objects for a particular file. This is because access to a particular file is not limited to a single process; a file can be opened concurrently by multiple processes. Because of this, the file object is used privately by every process.

The following are some of the differences in the way an inode and a file object are used by the kernel:

- File objects along with inodes are used when a process needs to access a file.

- To access the inode of the file, the process would need a file object pointing to the file inode. File objects belong to a single process, whereas an inode can be used by multiple processes.

- A file object is created whenever a file is to be opened. When another process wants to access the same file, a new file object, private to that process, will be created. Hence, we can say that a file object exists for every instance of an open file. But every file will always have a single inode.

- When a process closes the file, its corresponding file object is destroyed, but its inode might still be kept in the cache.

There can be some confusion between file objects and another similar entity that is used for accessing files, the **file descriptor**. An `open ()` system call by a process often returns a file descriptor, which is used by the process for accessing a file. In a way, a file descriptor also illustrates the relationship between a process and a file. So, what is the difference? To play with words, a file object provides an **open file description**. A file object will contain all data related to a file descriptor. File descriptors are user-space references to kernel objects. A file object will contain information such as a file pointer representing the current position in the file and how the file was opened:

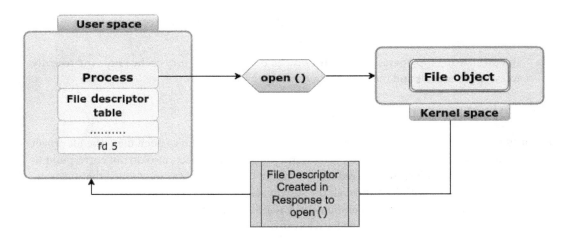

Figure 2.5 – A file object is created as a result of open ()

The definition for a file object is present in include/linux/fs.h. This structure stores information about the relationship of the process to the open file. The f_inode pointer points to the inode of the file:

```
struct file {
        union {
                struct llist_node        fu_llist;
                struct rcu_head          fu_rcuhead;
        } f_u;
        struct path                f_path;
        struct inode              *f_inode;          /* cached value */
        const struct file_operations    *f_op;
[.......]
```

Some important fields are described here:

- f_path: This field represents the directory path of the file associated with the open file. When a file is opened, the VFS creates a new struct file object and initializes its f_path field to point to the directory path of the file.

- f_inode: This is a pointer to the struct inode object that represents the file associated with the struct file object.

- f_op: This is a pointer to the struct file_operations object that contains a set of function pointers for file operations on the associated file.

- f_lock: This field is used to ensure synchronization between different threads that are accessing the same file object.

## Defining file operations

Like other structures, the filesystem methods applicable to file objects are defined in the file_
operations table. The f_op pointer is pointing to the file_operations table. The VFS
implements a common interface for all filesystems that hooks up with the actual mechanisms of the
underlying filesystems:

```
struct file_operations {
        struct module *owner;
        loff_t (*llseek) (struct file *, loff_t, int);
        ssize_t (*read) (struct file *, char __user *, size_t, loff_t
*);
        ssize_t (*write) (struct file *, const char __user *, size_t,
loff_t *);
        ssize_t (*read_iter) (struct kiocb *, struct iov_iter *);
        ssize_t (*write_iter) (struct kiocb *, struct iov_iter *);
        int (*iopoll)(struct kiocb *kiocb, struct io_comp_batch *,
unsigned int flags);
        int (*iterate) (struct file *, struct dir_context *);
        int (*iterate_shared) (struct file *, struct dir_context *);
        __poll_t (*poll) (struct file *, struct poll_table_struct *);
        long (*unlocked_ioctl) (struct file *, unsigned int, unsigned
long);
        long (*compat_ioctl) (struct file *, unsigned int, unsigned
long);
        int (*mmap) (struct file *, struct vm_area_struct *);
[.................]
```

The operations defined here look a lot similar to the generic system calls we described in *Chapter 1*:

- llseek: This is called when the VFS needs to move the file position index
- read: This is called by read() and related system calls
- open: This is called when a file (inode) needs to be opened
- write: This is called by write() and related system calls
- release: This is called when an open file is being closed
- map: This is called when a process wants to map a file in memory using the mmap() system call

These are generic operations. Not all of them can be applied to a single file. At the end of the day, it's
up to the individual filesystem to pick and choose from this set of operations. If a particular method
is not applicable to a filesystem, that can be simply set to NULL.

### Can a process run out of file descriptors?

The kernel enforces limits for the maximum number of processes that can be opened at a time. These limits can be applied at the user, group, or global system level. If all the file descriptors have been allocated, the process won't be able to open any more files. Many large applications require a lot more than the default number of descriptors allowed for a process. In such cases, the limits for individual users can be set in the `/etc/security/limits.conf` file. For system-wide settings, the `sysctl` command can be used:

```
[cyndaquil@linuxbox ~]$ touch file
-bash: start_pipeline: pgrp pipe: Too many open files
touch: error while loading shared libraries: libc.so.6: cannot open shared object file
[cyndaquil@linuxbox ~]$
```

Figure 2.6 – Too many open files spoil the application

Let's summarize our discussion before we reach our open file limit:

- A file object is an in-memory representation of an open file and does not have any corresponding on-disk image.

- A file object is created in response to an `open()` system call by a process.

- A file object is private for a process. As more than one process can access a particular file, the VFS will create multiple file objects for a particular file.

## Superblocks – describing filesystem metadata

If you've ever created a filesystem by running `mkfs` on a block device, chances are you must have seen the term **superblock** in the output. Superblock is one of the more familiar structures to the casual Linux user. You might have noticed that the structures used in VFS bear a close resemblance to each other. Dentry and file objects store in-memory representations of directories and open files, respectively. Both structures do not have an on-disk image and only exist in memory. On a similar note, the superblock structure has a lot in common with inodes. Inodes store metadata about files, whereas superblocks store metadata about filesystems.

Consider the example of a library catalog system that keeps track of the books, including their titles, authors, and locations on the shelves. If the catalog system is lost or damaged, it can be difficult to find and retrieve specific books in the library. Similarly, if the superblock structure in the kernel is corrupted or damaged, it can lead to data loss or filesystem errors.

Just as every file has an inode number assigned to it, every filesystem has a corresponding superblock structure. Like inodes, a superblock also has an on-disk image. For filesystems that generate their content on the fly, such as `procfs` and `sysfs`, their superblock structures are stored in memory only. When a filesystem is to be mounted, the superblock is the first structure that is read. Similarly, when the filesystem has been mounted, the information regarding the mounted filesystem is stored in the superblock.

The superblock of a filesystem contains intricate information about the filesystem, such as the total number of blocks, number of used, unused, and free blocks, filesystem state and type, inodes, and a lot more. As changes are made to the filesystem, the information stored in the superblock is updated. Since the superblock is read while mounting a filesystem, we have to ask what would happen if the information stored in the superblock gets erased or corrupted. To put it simply, a filesystem cannot be mounted without a superblock. Given its critical nature, several copies of the superblock are saved in multiple disk locations. In the case that the primary superblock is corrupted, a filesystem can be mounted using any of the backup superblocks.

The superblock structure is defined in `include/linux/fs.h`. The `s_list` contains pointers to mounted superblocks and `s_dev` identifies the device. The superblock operations are defined in the `super_operations` table, pointed at by the `s_op` pointer:

```
struct super_block {
        struct list_head        s_list;         /* Keep this first */
        dev_t                   s_dev;          /* search index; _not_
kdev_t */
        unsigned char           s_blocksize_bits;
        unsigned long           s_blocksize;
        loff_t                  s_maxbytes;     /* Max file size */
        struct file_system_type *s_type;
        const struct super_operations   *s_op;
        const struct dquot_operations   *dq_op;
        const struct quotactl_ops       *s_qcop;
        const struct export_operations *s_export_op;
[..............................]
```

Some important fields of the superblock structure are explained as follows:

- `s_list`: This field is used to maintain the list of all the currently mounted filesystems.

- `s_dev`: This field specifies the device number that corresponds to the filesystem's root directory inode. This is used to identify the device on which the filesystem resides.

- `s_type`: This field points to the definition of the specific filesystem that is used to interpret the data stored on the filesystem. For instance, if this points to the XFS filesystem, the kernel knows that it needs to use XFS-specific functions to interact with the filesystem.

- `s_root`: This field is used by the kernel to locate the root directory of the filesystem when it is mounted. Once the root directory has been identified, the directory tree can be traversed to access the other files and directories in the filesystem.

- `s_magic`: This field is used to identify the filesystem type on a particular device or partition.

Again, there are a considerable number of fields, so it's not possible to explain all of them. Some fields are simple integers, while others have far more complex data structures and function pointers.

## Superblock operations in the kernel

As with all VFS structures, all the superblock operations in `include/linux/fs.h` are not mandatory for a filesystem. The kernel keeps a copy of the filesystem superblock in the memory. When changes are made in the filesystem, the information in the superblock is updated in memory. The superblock copy in memory is thus marked `dirty` as the kernel needs to update the on-disk superblock with the updated information:

```
struct super_operations {
        struct inode *(*alloc_inode)(struct super_block *sb);
        void (*destroy_inode)(struct inode *);
        void (*free_inode)(struct inode *);
        void (*dirty_inode) (struct inode *, int flags);
        int (*write_inode) (struct inode *, struct writeback_control
*wbc);
        int (*drop_inode) (struct inode *);
        void (*evict_inode) (struct inode *);
        void (*put_super) (struct super_block *);
        int (*sync_fs)(struct super_block *sb, int wait);
        int (*freeze_super) (struct super_block *);
[.............................]
```

Some important methods are defined as follows:

- `alloc_inode`: This method is called to initialize and allocate memory for `struct inode`
- `destroy_inode`: This method is called by `destroy_inode()` to release resources allocated for `struct inode`
- `dirty_inode`: This method is called by the VFS to mark an inode as `dirty`
- `write_inode`: This method is called when the VFS needs to write an inode to disk
- `delete_inode`: This is called when the VFS wants to delete an inode
- `sync_fs`: This is called when VFS is writing out all `dirty` data associated with a superblock
- `statfs`: This is called when the VFS needs to get filesystem statistics, such as its size, free space, and number of inodes
- `umount_begin`: This is called when the VFS is unmounting a filesystem

Let's summarize:

- The superblock structure records all filesystem characteristics.
- The superblock structure is read when mounting and unmounting a filesystem. Filesystems maintain copies of superblocks in multiple disk locations.

## Linking

In our discussion regarding directory entries, we mentioned the **linking** operation. Links are of two types: symbolic (or soft) links and hard links, as most users would know. Symbolic (soft) links behave as shortcuts, although there are subtle differences. **Soft links** point to the path that contains the data, while **hard links** refer to the data itself.

Going back a bit, the inode doesn't contain the name of the file. The name of the file is contained within the directory. That means there can be multiple filenames in a directory list, all of which point to the same inode. Hard links use this logic. A hard link points to the inode of the file. This means that the link and file are indistinguishable as both are pointing to the same inode. After a while, you might not even know which was the original file:

```
[root@linuxbox ~]# ls -li file1.txt
134788052 -rw-r--r-- 2 root root 246309 Oct  3 10:26 file1.txt
[root@linuxbox ~]#
[root@linuxbox ~]# ls -li /tmp/file1.txt
134788052 -rw-r--r-- 2 root root 246309 Oct  3 10:26 /tmp/file1.txt
[root@linuxbox ~]#
```

Figure 2.7 – It's impossible to tell which is the original file as both have the same inode

In contrast, a symbolic link has a different inode number than the original file. Note how this symbolic link is pointing to the original file and indicates the l in the permission section of the file:

```
[root@linuxbox ~]# ls -li /tmp/file2.txt
202131972 lrwxrwxrwx 1 root root 15 Oct 13 04:07 /tmp/file2.txt -> /root/file2.txt
[root@linuxbox ~]#
[root@linuxbox ~]# ls -li /root/file2.txt
134460859 -rw-r--r-- 1 root root 246309 Oct 13 04:07 /root/file2.txt
[root@linuxbox ~]#
```

Figure 2.8 – For soft links, the first character in the file permissions is "l"

Using the same inode number for multiple files results in some limitations. As inode numbers are only unique within a filesystem, hard links cannot span across filesystem boundaries. They can only exist within a filesystem. Hard links can only be used for regular files, not directories. This is to prevent breaking the filesystem structure. A hard link to a directory could create an endless loop structure.

## Summarizing the four structures

Some of the concepts that we've discussed here might become a lot more clear when we discuss block filesystems in *Chapter 3, Exploring the Actual Filesystems Under the VFS*. But we have gotten some understanding of how VFS goes about spinning that web of abstraction.

As discussed in *Chapter 1*, the design of VFS is biased toward filesystems that originate from the Linux tribe. Most non-native filesystems do not speak in terms of inodes, superblocks, files, and directory objects. To implement the common file model for them, VFS creates these structures in memory. So, objects such as inodes and superblocks, which have an on-disk and in-memory presence for native filesystems, might only be present in memory for non-native filesystems. Because of the difference in the design of non-native filesystems, they might not support some common filesystem operations in Linux, such as symbolic links.

The following table provides a brief summary of the major VFS structures:

| Structure | Description | Stored on disk/in memory |
|---|---|---|
| Inode | Contains all file metadata except the filename | On disk and in memory |
| Dentry | Represents the relationship between a directory entry and files | Only in memory |
| File Object | Stores information about the relationship of the process to an open file | Only in memory |
| Superblock | Holds filesystem characteristics and metadata | On disk and in memory |

Table 2.1 – Summarizing major VFS data structures

The following figure represents how a process will go about opening a file stored on disk:

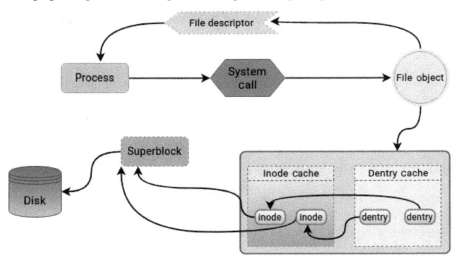

Figure 2.9 – Relationship between common VFS structures

Remember that the superblock structure is created and initialized during the filesystem mount process and it contains a pointer to the root dentry, which, in turn, contains a pointer to the inode that represents the root directory of the filesystem.

When a process calls the `open()` system call to open a file, the VFS creates a `struct file` object to represent the file in the process's address space and initializes its `f_path` field to point to the directory path of the file. The `struct dentry` object contains a pointer to a `struct inode` object, which represents the on-disk inode for the file.

The `struct inode` object is associated with the `struct super_block` object, which represents the on-disk filesystem. The `struct super_block` object contains pointers to the filesystem-specific functions defined in the `struct super_operations` structure, which is used by the VFS to interact with the filesystem.

## Page cache

The nomenclature used for defining different concepts and terms in Linux is a bit odd. The system call for creating a file is known as `creat`. Ken Thompson, the creator of Unix, once jokingly said that the missing *e* in `creat` was his greatest regret in the Unix design. While explaining a few operations of VFS structures, the word *dirty* has been used. How and why this term has been used in Linux is anybody's guess. The term *dirty* here refers to pages in memory that have been modified but have not yet been *flushed* to disk.

Caching is a common practice used in both hardware and software applications to improve performance. In terms of hardware, the speed and performance of the CPU, memory subsystem, and physical disks are interconnected. The CPU is much faster than the memory subsystem, which, in turn, is faster than physical disks. This discrepancy in speed can result in wasted CPU cycles when waiting for a response from memory or disk.

To address this issue, cache layers were added to the CPU to store frequently accessed data from the main memory. This allows the CPU to operate at its natural speed as long as the required data is available in the cache. Similarly, software applications also use caching to store frequently accessed data or instructions in a faster and more accessible location to improve performance.

The Linux design is geared toward performance, and the page cache plays a vital role in ensuring this. The primary purpose of the page cache is to improve the latency of `read` and `write` operations by ensuring that data is kept in memory (provided there is enough available) so that frequent trips to the underlying physical disks can be avoided. All this contributes toward performance improvement as disk access is far slower than memory.

Operating systems interact with hardware at a lower level and use different units for managing and utilizing the available resources. For example, filesystems break down the disk space into blocks, which is a higher-level abstraction than individual bytes or bits. The reason for this is that managing data at a byte or bit level can be complex and time-consuming. A page is the fundamental unit of memory in the kernel and is 4 KB by default. This is significant because all I/O operations are aligned to some number of pages. The following figure summarizes how data is read from and written to disk, using the page cache:

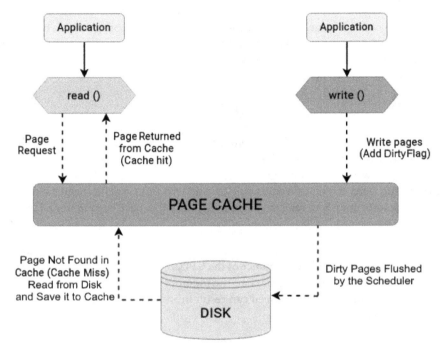

Figure 2.10 – Improving I/O performance through Page Cache

*Figure 2.10* highlights how the page cache aids in improving read and write performance by caching frequently accessed file data in memory, thereby reducing disk access and improving system performance.

## Reading

The following points provide a brief summary of what happens when a process running in user space requests data to be read from disk:

1.   The kernel first checks whether the required data is already available in the cache. If the data is found in the cache, the kernel can avoid performing any disk operation and directly provide the requested data to the process. This scenario is known as a **cache hit**.

2.   If the requested data cannot be found in the cache, the kernel must go to the underlying disk. This is known as a **cache miss**. Before that, it has to check whether any free memory is available. The kernel will then schedule a `read` operation, retrieve the requested data from the disk, save it to the cache, and then hand it over to the calling process.

3.   If any further requests are made to access this page, they can be fulfilled from the page cache.

4.   In the case where the requested data is found in the cache but has already been marked as `dirty`, the kernel will first write it back to disk before proceeding with the procedure mentioned earlier.

Similarly, when a process needs to write data to disk, the following happens:

1.   The kernel updates the page cache that is mapped to the file and marks the data as `dirty`. The pages that are yet to be written to the disk are known as **dirty pages**.

2.   The kernel will not immediately write all dirty data to disk. Depending upon the configuration of the kernel's flusher thread, the dirty data will be flushed to the disk.

After a `write` request is completed, the kernel sends an acknowledgment to the calling process. However, it doesn't inform the process about when the corresponding dirty page will actually be written to disk. It is interesting to note that this asynchronous approach makes write operations much faster than read operations, as the kernel dodges a trip to the underlying physical disk. This also begs the question: *If my I/O requests are served from memory, what happens to my data in case of abrupt power loss?* The dirty pages in memory do get flushed to the physical disks after a certain interval; this is called **writeback**. How often or how long data remains in memory depends on certain factors. If an application keeps on caching data, it can create problems for other processes in the system. The amount of total memory in a system is far less than the physical disk capacity, so it makes sense to clear up the cache after regular intervals. In order to prevent the loss of critical data, some applications, such as databases, need guarantees that data has been written to the persistent storage. In such cases, it makes sense to flush the dirty data immediately to disk. The kernel offers some **tunables** (through `sysctl`), which can be used for controlling this behavior of the page cache.

Despite the risks associated with page caching, there is no doubt that it improves performance. The size of the page cache is not fixed; it is dynamic in nature. The page cache can use the available memory resources. However, when the free memory available in the system goes below a threshold, the flushing schedulers will kick in and start to unload data from the page cache onto disk.

# Summary

The flexibility of support for Linux filesystems is a direct result of an abstracted set of interfaces implemented by the VFS. In this chapter, we learned about the major data structures in the VFS and how they are all working together. The VFS uses several data structures to implement generic abstraction methods for the different native and non-native filesystems. The four most common structures are inodes, directory entries, file objects, and superblocks. These structures ensure commonality between the design and operations of different filesystems. Since the methods defined by the VFS are generic, it is not compulsory for filesystems to implement all of them, although the filesystems should adhere to the structures defined in the VFS and build upon them to ensure a generic interface is maintained.

In addition to filesystem abstractions, the VFS also provides a number of caches to improve the performance of filesystem operations, such as dentry and inode cache. We also explained the mechanism of the page cache in the kernel and saw how it can speed up read and write requests issued by the user-space programs. In *Chapter 3*, we're going to explore the actual filesystems under the VFS layer. We'll cover some popular Linux filesystems, primarily the extended filesystem and how it organizes user data on disk. We'll also explain some common concepts associated with the different filesystems in Linux, such as journaling, copy-on-write, and filesystems in user space.

# 3

# Exploring the Actual Filesystems Under the VFS

*"Not all roots are buried down in the ground, some are at the top of a tree." — Jinvirle*

The kernel's I/O stack can be broken down into three major sections: the **virtual filesystem** (**VFS**), the **block layer**, and the **physical layer**. The different flavors of filesystems supported by Linux can be thought of as the tail end of the VFS layer. The first two chapters gave us a decent understanding of the role of VFS, the major structures used by VFS, and how it aids the end user processes to interact with the different filesystems through a common file model. This means that we'll now be able to use the word *filesystem* in its commonly accepted context. Finally.

In *Chapter 2*, we defined and explained some important data structures used by the VFS to define a generic framework for different filesystems. In order for a particular filesystem to be supported by the kernel, it should operate within the boundaries defined in this framework. But it is not mandatory that all the methods defined by the VFS are used by a filesystem. The filesystems should stick to the structures defined in the VFS and build upon them to ensure commonality between them, but as each filesystem follows a different approach for organizing data, there might be a ton of methods and fields in these structures that are not applicable to a particular filesystem. In such cases, filesystems define the relevant fields as per their design and leave out the non-essential information.

As we've seen, the VFS is sandwiched between user-space programs and *actual* filesystems and implements a common file model so that applications can use uniform access methods to perform their operations, regardless of the underlying filesystem in use. We're now going to shift our focus to one particular side of this *sandwich*, which is the filesystems that contain user data.

This chapter will introduce you to some of the more common and popular filesystems used in Linux. We'll cover the working of the extended filesystem in great detail as it is most commonly used. We'll also shed some light on **network filesystems**, and cover a few important concepts related to filesystems such as journaling, filesystems in user-space, and **copy-on-write** (**CoW**) mechanisms.

We're going to cover the following main topics:

- The concept of journaling
- CoW mechanisms
- The extended filesystem family
- Network filesystems
- Filesystems in user space

## Technical requirements

This chapter focuses entirely on filesystems and associated concepts. If you have experience with storage administration tasks in Linux but haven't delved into the inner workings of filesystems, this chapter will serve as a valuable exercise. Having prior knowledge of filesystem concepts will enhance your understanding of the content covered in this chapter.

The commands and examples presented in this chapter are distribution-agnostic and can be run on any Linux operating system, such as Debian, Ubuntu, Red Hat, Fedora, and so on. There are a few references to the kernel source code. If you want to download the kernel source, you can download it from `https://www.kernel.org`. The code segments referred to in this book are from kernel `5.19.9`.

## The Linux filesystem gallery

As said earlier, one of the major benefits of using Linux is the wide range of supported filesystems. The kernel contains out-of-the-box support for some of these, such as XFS, Btrfs, and extended filesystem versions 2, 3, and 4. These are considered **native filesystems** as they were designed keeping in mind the Linux principles and philosophies. On the other side of the aisle are filesystems such as NTFS and FAT. These can be considered **non-native filesystems**. This is because, although the Linux kernel is capable of understanding these filesystems, supporting them usually requires additional configuration as they do not fall in line with the conventions adopted by native filesystems. We're going to keep our focus on the native filesystems and explain the key concepts associated with them.

Although each filesystem claims to be better, faster, and more reliable and secure than all others, it is important to note that no filesystem can be the best fit for all kinds of applications. Every filesystem comes with its strengths and limitations. From a functional standpoint, filesystems can be classified as follows:

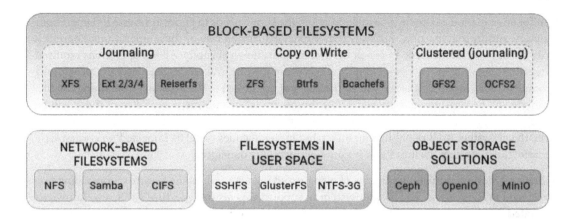

Figure 3.1 – Linux filesystem gallery

*Figure 3.1* gives a glimpse of some of the supported filesystems and their respective categories. Given the plethora of filesystems supported by Linux, covering all of them will make us run out of space (filesystem pun!). Although the implementation details vary, filesystems usually make use of some common techniques for their internal operations. Some core concepts, such as journaling, are more common among filesystems. Similarly, some filesystems make use of the popular CoW technique, due to which they do not need journaling.

Let's explain the concept of journaling in filesystems.

## The diary of a filesystem – the concept of journaling

A filesystem uses complex structures to organize data on the physical disk. In the case of a system crash or abrupt failure, a filesystem is unable to finish off its operations in a graceful manner, which can corrupt its organizational structures. When the system is powered up the next time, the user will need to run a consistency or integrity check of some sort against the filesystem to detect and repair those damaged structures.

When explaining VFS data structures in *Chapter 2*, we discussed that one of the fundamental principles followed in Linux is the separation of metadata from actual data. The metadata of a file is defined in an independent structure, called an **inode**. We also saw how a directory is treated as a special file and it contains the mapping of filenames to their inode numbers. Keeping this in mind, let's say we're creating a simple file to add some text to it. To go through with this, the kernel will need to perform the following operations:

1.  Create and initialize a new inode for the file to be created. An inode should be unique within a filesystem.

2.  Update the timestamps for the directory in which the file is being created.

3.  Update the inode for the directory. This is required so that the filenames-to-inode mapping is updated.

Even for an operation as simple as text file creation, the kernel needs to perform several I/O operations to update multiple structures. Let's say that while performing one of these operations, there is a hardware or power failure due to which the system shuts down abruptly. All the operations required for creating a new file will not have completed successfully, which will render the filesystem structurally incomplete. If an inode for the file was initialized and not linked to the directory containing the file, the inode will be considered *orphaned*. Once the system is back online, a consistency check will be run on the filesystem, which will remove any such inodes that are not linked to any directory. After a crash, the filesystem itself might remain intact, but individual files could be impacted. In a worst-case scenario, the filesystem itself can also become permanently damaged.

To improve filesystem reliability in case of outages and system crashes, the feature of journaling was introduced in filesystems. The first filesystem to support this feature was IBM's **JFS**, also known as **Journaled Filesystem**. Over the last few years, journaling has become an essential ingredient in the design of filesystems.

The concept of filesystem journaling finds its roots in the design of database systems. In most databases, journaling guarantees data consistency and integrity in case a transaction fails due to external events, such as hardware failures. A database journal will keep track of uncommitted changes by recording such operations in a journal. When the system comes back online, the database will perform a recovery using the journal. Journaling in filesystems follows the same route.

Any changes that need to be performed on the filesystem are first written sequentially to a journal. These changes or modifications are referred to as transactions. Once a transaction has been written to a journal, it is then written to the appropriate location on the disk. In the case of a system crash, the filesystem replays the journal to see whether any transaction is incomplete. When the transaction has been written to its on-disk location, it is then removed from the journal.

Depending on the journaling approach, either metadata or actual data (or both) is first written to the journal. Once data has been written to the filesystem, the transaction is removed from the journal:

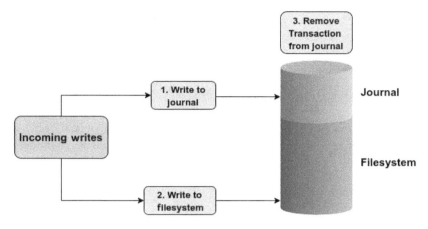

Figure 3.2 – Journaling in filesystems

It's important to note that, by default, the filesystem journal is also stored on the same filesystem, albeit in a segregated section. Some filesystems also allow storing the journal on a separate disk. The size of the journal is typically just a few megabytes.

## The burning question – doesn't journaling adversely affect performance?

The entire point of journaling is to make the filesystem more reliable and preserve its structures in case of system crashes and hardware failures. With a journaling filesystem, data is first written to a journal and then to its specified disk location. It's not difficult to see that we're adding an extra hop to reach our destination as we'll need to write the same data twice. Surely this is going to backfire and undermine the filesystem performance?

It's one of those questions whose answer seems obvious but it isn't. The filesystem performance doesn't necessarily deteriorate when using journaling. In fact, in most cases, it's the exact opposite. There could be workloads where the difference in both cases is negligible, but in most scenarios, especially in metadata-intensive workloads, filesystem journaling actually boosts performance. The degree to which the performance is enhanced can vary.

Consider a filesystem without journaling. Every time a file is modified, the natural course of action is to perform the relevant modifications on the disk. For metadata-intensive operations, this could negatively impact performance. For instance, modifications in file contents also require that the corresponding timestamps of the file are also updated. This means that every time a file is processed and modified, the filesystem has to go and update not only the actual file data but also the metadata. When journaling is enabled, **fewer seeks** to the physical disk are required as data is written to disk only when a transaction has been committed to the journal or when the journal fills up. Another benefit is the use of sequential writes in a journal. When using a journal, random write operations are converted into sequential writes.

In most cases, performance improvement occurs as a result of the cancellation of metadata operations. When metadata updates are required in a swift manner, such as recursively performing operations on a directory and its contents, the use of journaling can improve performance by reducing frequent trips to disks and performing multiple updates in an atomic operation.

Of course, how a filesystem implements journaling also plays a major role in this. Filesystems offer different approaches when it comes to journaling. For instance, some filesystems only journal the metadata of a file, while others write both metadata and actual data in a journal. Some filesystems also offer flexibility in their approach and allow end users to decide the journaling mode.

To summarize, journaling is an important constituent of modern filesystems as it makes sure that the filesystem remains structurally sound, even in the case of a system crash.

## The curious case of CoW filesystems

**CoW** is a resource management mechanism that is used in the Linux kernel. This concept is most commonly associated with the `fork ()` system call. The `fork ()` system call creates a new process by duplicating the calling process. When a new process is created using a `fork ()` system call, memory pages are shared between the parent and child processes. As long as the pages are being shared, they cannot be modified. When either the parent or child process attempts to modify a page, the kernel duplicates the page and marks it writable.

Most filesystems in Linux that have existed for a long time use a very conventional approach when it comes to the core design principles. Over the past several years, two major changes in the extended filesystem have been the use of journaling and extents. Although efforts have been made to scale the filesystems for modern use, some major areas such as error detection, snapshots, and deduplication have been left out. These features are the need of today's enterprise storage environments.

Filesystems that use the CoW approach for writing data differ from other filesystems in a notable way. When overwriting data on an Ext4 or XFS filesystem, the new data is written on top of the existing data. This means that the original data will be destroyed. Filesystems that use the CoW approach copy the old data to some other location on disk. The new incoming data is written to this new location. Hence, the phrase *Copy on Write*. As the old data or its snapshot is still there, the space utilization on the filesystem will be a lot more than the user would expect to see. This is often confusing to newer users and it might take some time to get used to this. Some Linux folks have a rather funny take on this: *CoWs ate my data*. As shown in *Figure 3.3*, filesystems using the CoW approach write incoming data to a new block:

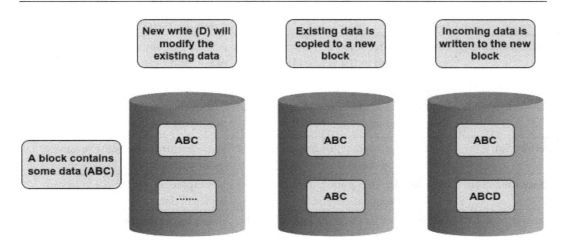

Figure 3.3 – The CoW approach in filesystems

As an analogy, we can loosely relate this to the concept of time travel in movies. When someone travels back in time and makes changes to the past, a parallel timeline is created. This creates a separate copy of the timeline that diverges from the original. CoW filesystems operate similarly. When a modification is requested on a file, instead of directly modifying the original data, a separate copy of the data is created. The original data remains intact while the modified version is stored separately.

Since the original data is preserved in the process, this opens up some interesting avenues. Because of this approach, filesystem recovery in the case of a system crash is simplified. The previous state of data is saved on an alternate location on disk. Hence, if there's an outage, the filesystem can easily revert to its former state. This makes the need for maintaining any journal obsolete. This also allows for the implementation of snapshots at the filesystem level. Only modified data blocks are copied to a new location. When a filesystem needs to be restored using a particular snapshot, the data can be easily reconstructed.

*Table 3.1* highlights some major differences between journaling and CoW-based filesystems. Please note that the implementation and availability of some of these features may vary depending on the type of filesystem:

|  | Journaling | Copy-on-Write |
|---|---|---|
| Write handling | Changes are recorded in a journal before applying them to the actual filesystem | A separate copy of data is created to make modifications |
| Original data | Original data gets overwritten | Original data remains intact |
| Data consistency | Ensures consistency by recording metadata changes and replaying them if needed | Ensures consistency by never modifying the original data |
| Performance | Minimal overhead depending on the type of journaling mode | Some performance gains because of faster writes |
| Space utilization | Journal size is typically in MB, so no additional space is required | More space is required due to separate copies of data |
| Recovery times | Fast recovery times as the journal can be replayed instantly | Slower recovery times as data needs to be reconstructed using recent copies |
| Features | No built-in support for features such as compression or deduplication | Built-in support for compression and deduplication |

Table 3.1 – Differences between CoW and journaling filesystems

Filesystems that use the CoW-based approach for organizing data include **Zettabyte Filesystem (ZFS)**, **B-Tree Filesystem (Btrfs)**, and Bcachefs. ZFS was initially used on Solaris and quickly gained popularity because of its powerful features. Although not included in the kernel because of licensing issues, it has been ported to Linux through the *ZFS on Linux* project. The Bcachefs filesystem was developed from the kernel's block cache code and is quickly gaining popularity. It might become a part of future kernel releases. Btrfs, also fondly known as ButterFS, is directly inspired by ZFS. Unfortunately, because of a few bugs in early releases, its adoption slowed down in the Linux community. Nevertheless, it has been under active development and has been a part of the Linux kernel for over a decade.

Despite a few issues, Btrfs is the most advanced filesystem present in the kernel because of its rich feature set. As mentioned previously, Btrfs draws a lot of inspiration from ZFS and tries to offer almost identical features. Like ZFS, Btrfs is not just a simple disk filesystem, it also offers the functionality of a logical volume manager and software, **Redundant Array of Independent Disks (RAID)**. Some of its features include snapshots, checksums, encryption, deduplication, and compression, which are usually not available in regular block filesystems. All these characteristics greatly simplify storage management.

To summarize, the CoW approach of filesystems such as Btrfs and ZFS ensures that existing data is never overwritten. Hence, even in the case of a sudden system crash, existing data will not be in an inconsistent state.

# Extended filesystem

The **extended filesystem**, shortened as **Ext**, has been a trusted aide of the Linux kernel since its inception and is almost as old as the Linux kernel itself. It was first introduced in the kernel 0.96c. Over the years, the extended filesystem has gone through some major changes that have resulted in multiple versions of the filesystem. These versions are briefly explained as follows:

- **The First Extended Filesystem**: The first filesystem to run Linux was Minix and it supported a maximum filesystem size of 64 MB. The extended filesystem was designed to overcome the shortcomings in Minix and was generally considered an extension of the Minix filesystem. The extended filesystem supported a maximum filesystem size of 2 GB. It was also the first filesystem to make use of the VFS. The first Ext filesystem only allowed one timestamp per file, as compared to the three timestamps used today.

- **The Second Extended Filesystem**: Almost a year after the release of the first extended filesystem, its second version, Ext2, was released. The Ext2 filesystem addressed the limitations of its predecessor, such as partition sizes, fragmentation, filename lengths, timestamps, and maximum file size. It also introduced several new features including the concept of filesystem blocks. The design of Ext2 was inspired by BSD's Berkeley Fast File System. The Ext2 filesystem supported much larger filesystem sizes, up to a few terabytes.

- **The Third Extended Filesystem**: The Ext2 filesystem was widely adopted but fragmentation and filesystem corruption in the case of a crash remained big concerns. The third extended filesystem, Ext3, was designed keeping this in mind. The most important feature introduced in this release was **journaling**. Through journaling, the Ext3 filesystem kept track of uncommitted changes. This reduced the risk of data loss if the system crashed because of a hardware or power failure.

- **The Fourth Extended Filesystem**: Ext4 is currently the latest version of the extended filesystem family. The Ext4 filesystem offers several improvements over Ext2 and Ext3 in terms of performance, fragmentation, and scalability, while also keeping backward compatibility with Ext2 and Ext3. When it comes to Linux distributions, Ext4 is probably the most frequently deployed filesystem.

We're going to mainly focus on the design and structure of the most recent version of the extended filesystem, Ext4.

## Blocks – the lingua franca of filesystems

At the lowest level, a hard drive is addressed in units of sectors. Sectors are the physical property of a disk drive and are normally 512 bytes in size. Although, these days, it's not uncommon to see drives using a sector size of 4 KB. The sector size is something that we cannot tinker with as it is decided by the drive manufacturer. As a sector is the smallest addressable unit on the drive, any operation performed on the physical drive is always going to be larger than or equal to the sector size.

A filesystem is created on top of the physical drive and does not address the drive in terms of sectors. All filesystems (and the extended filesystem family is no exception to this) address a physical drive in terms of blocks. A block is a group of physical sectors and is the fundamental unit of a filesystem. An Ext4 filesystem performs all operations in terms of blocks. On x86 systems, the filesystem block size is set to 4 KB by default. Although it can be set to a lower or higher value, the block size should always satisfy the following two constraints:

- The block size should always be a power-of-two multiple of the disk sector size
- The block size should always be less than or equal to the memory page size

The maximum filesystem block size is the page size of the architecture. On most x86-based systems, the default page size of the kernel is 4 KB. So, the filesystem block size cannot exceed 4 KB. The page size of the VFS cache also amounts to 4 KB. The restriction of the block size to be less than or equal to the kernel's page size is not limited to the extended filesystem only. The page size is defined during kernel compilation and is 4 KB for x86_64 systems. As shown ahead, the mkfs program for Ext4 will throw a warning if a block size greater than the page size is specified. Even if a filesystem is created with a block size greater than the page size, it cannot be mounted:

```
[root@linuxbox ~]# getconf PAGE_SIZE
4096
[root@linuxbox ~]# mkfs.ext4 /dev/sdb -b 8192
Warning: blocksize 8192 not usable on most systems.
mke2fs 1.44.6 (5-Mar-2019)
mkfs.ext4: 8192-byte blocks too big for system (max 4096)
Proceed anyway? (y,N) y
Warning: 8192-byte blocks too big for system (max 4096), forced to
continue
[....]
[root@linuxbox ~]# mount /dev/sdb /mnt
mount: /mnt: wrong fs type, bad option, bad superblock on /dev/sdb,
missing codepage or helper program, or other error.
[root@linuxbox ~]#
[root@linuxbox ~]# dmesg |grep bad
[ 5436.033828] EXT4-fs (sdb): bad block size 8192
[ 5512.534352] EXT4-fs (sdb): bad block size 8192
[root@linuxbox ~]#
```

Once a filesystem has been created, the block size cannot be changed. The Ext4 filesystem divides the available storage into logical blocks of 4 KB by default. The selection of block size has a significant impact on the space efficiency and performance of the filesystem. The block size dictates the minimum on-disk size of a file, even if its actual size is less than the block size. Let's say that our filesystem uses a block size of 4 KB and we save a simple text file of 10 bytes on it. This 10-byte file, when stored on the physical disk, will use 4 KB of space. A block can only hold a single file. This means that for a 10-byte file, the remaining space in a block (4 KB – 10 bytes) is wasted. As shown next, a simple text file containing the "hello" string will occupy a full filesystem block:

```
robocop@linuxbox:~$ echo "hello" > file.txt
robocop@linuxbox:~$ stat file.txt
  File: file.txt
  Size: 6              Blocks: 8          IO Block: 4096    regular
file
Device: 803h/2051d     Inode: 2622288     Links: 1
Access: (0664/-rw-rw-r--)  Uid: ( 1000/   robocop)  Gid: (
1000/   robocop)
Access: 2022-11-10 12:55:55.406596713 +0500
Modify: 2022-11-10 13:01:12.962761327 +0500
Change: 2022-11-10 13:01:12.962761327 +0500
  Birth: -
robocop@linuxbox:~$
```

The stat command gives us a block count of 8, which is a bit misleading, as it is actually the sector count. This is because the stat system call assumes that 512 bytes of disk space are allocated per block. The block count here indicates that 4096 bytes (8 x 512) are physically allocated on the disk. The file size is 6 bytes only, but it occupies one full block on the disk. As shown next, when we add another line of text in the file, the file size increases from 6 to 19 bytes, but the numbers of used sectors and blocks remain the same:

```
robocop@linuxbox:~$ echo "another line" >> file.txt
robocop@linuxbox:~$ stat file.txt
  File: file.txt
  Size: 19             Blocks: 8          IO Block: 4096    regular
file
Device: 803h/2051d     Inode: 2622288     Links: 1
Access: (0664/-rw-rw-r--)  Uid: ( 1000/   robocop)  Gid: (
1000/   robocop)
Access: 2022-11-10 12:55:55.406596713 +0500
Modify: 2022-11-10 13:01:59.772249416 +0500
Change: 2022-11-10 13:01:59.772249416 +0500
  Birth: -
robocop@linuxbox:~$
```

## Is there a more efficient approach to organizing data?

Given that a small text file occupies a full block, it's not difficult to see the impact of filesystem block size. Having a lot of small files on a filesystem of a large block size can result in a waste of disk space, and a filesystem can quickly run out of blocks. Let's see a visual representation for a clearer understanding.

Let's say we have four files of varied sizes as follows:

- File A -> 5 KB
- File B -> 1 KB
- File C -> 7 KB
- File D -> 2 KB

Going with the approach of allocating a whole block (4 KB) to a single file, the files will be stored on disk as follows:

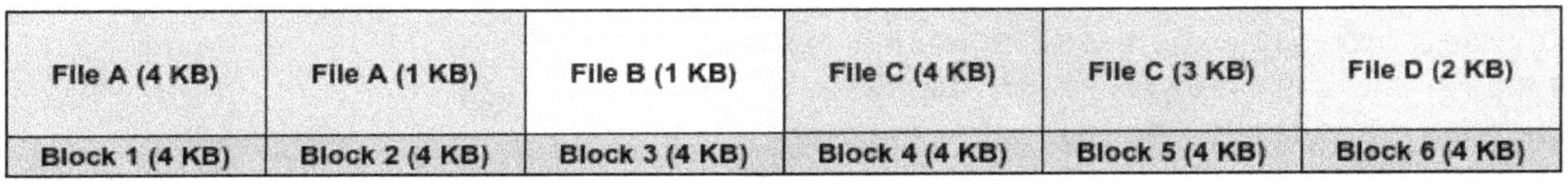

| File A (4 KB) | File A (1 KB) | File B (1 KB) | File C (4 KB) | File C (3 KB) | File D (2 KB) |
|---------------|---------------|---------------|---------------|---------------|---------------|
| Block 1 (4 KB) | Block 2 (4 KB) | Block 3 (4 KB) | Block 4 (4 KB) | Block 5 (4 KB) | Block 6 (4 KB) |

Figure 3.4 – Even the smallest of files occupy a full block

As is apparent from *Figure 3.4*, we're wasting 3 KB of space in blocks 2 and 3, and 1 KB and 2 KB in blocks 5 and 6, respectively. It's pretty clear that too many small files spoil the blocks!

Let's try an alternative approach and try to store the files in a more condensed format to avoid wasting space:

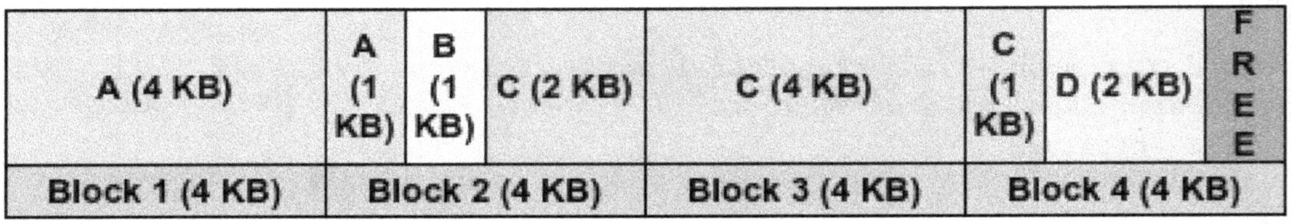

| A (4 KB) | A (1 KB) | B (1 KB) | C (2 KB) | C (4 KB) | C (1 KB) | D (2 KB) | FREE |
|----------|----------|----------|----------|----------|----------|----------|------|
| Block 1 (4 KB) | | Block 2 (4 KB) | | Block 3 (4 KB) | | Block 4 (4 KB) | |

Figure 3.5 – An alternate method for storing files

It's not difficult to see that the second approach is more compact and efficient. We're now able to store the same four files in only four blocks as compared to the six in the first approach. We are even able to save 1 KB of filesystem space. Apparently, allocating a whole filesystem block for a single file seems like an inefficient method for managing space, but in reality, that is a necessary evil.

On the first look, the second approach seems far better, but do you see the design flaw? A filesystem following this approach would have major pitfalls. If filesystems were designed to accommodate multiple files in a single block, they would need to devise a mechanism that would keep track of individual file boundaries within a single block. This would increase the design complexity by a fair margin. Additionally, this would lead to massive fragmentation, which would degrade the filesystem performance. If the size of a file increases, the incoming data would have to be adjusted in a separate block. Files would be stored in random blocks and there would be no sequential access. All of this would result in poor filesystem performance and neutralize any advantage gained from this condensed approach. Therefore, every file occupies a full block, even if its size is less than the filesystem block size.

## The structural layout of an Ext4 filesystem

The individual blocks in Ext4 are arranged into another unit called block groups. A **block group** is a collection of contiguous blocks. When it comes to the organization of the block group, there are two cases. For the first block group, the first 1,024 bytes are not used. These are reserved for the installation of boot sectors. For the first block group, the layout is as follows:

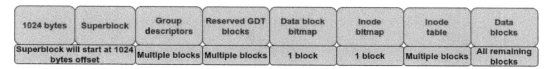

Figure 3.6 – Layout for block group 0

If the filesystem is created with a block size of 1 KB, the superblock will be kept in the next block. For all other block groups, the layout becomes the following:

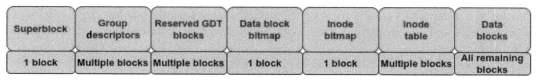

Figure 3.7 – Layout for block group 1 and onward

Let's discuss the constituents of an Ext4 block group.

## Superblock

As explained in *Chapter 2*, the **superblock** is one of the major data structures in VFS. It is mandatory for a filesystem to implement a superblock structure that contains the metadata of a filesystem. The Ext4 superblock is defined in `fs/ext4/ext4.h` and, as shown next, it contains dozens of fields defining the different attributes of the filesystem:

```
struct ext4_super_block {
    __le32   s_inodes_count;            /* Inodes count */
        __le32   s_blocks_count_lo;        /* Blocks count */
        __le32   s_r_blocks_count_lo;      /* Reserved blocks count */
        __le32   s_free_blocks_count_lo;   /* Free blocks count */
    __le32   s_free_inodes_count;       /* Free inodes count */
        __le32   s_first_data_block;       /* First Data Block */
        __le32   s_log_block_size;         /* Block size */
        __le32   s_log_cluster_size;       /* Allocation cluster size */
/*20*/  __le32   s_blocks_per_group;       /* # Blocks per group */
        __le32   s_clusters_per_group;     /* # Clusters per group */
        __le32   s_inodes_per_group;       /* # Inodes per group */
        __le32   s_mtime;                  /* Mount time */
/*30*/  __le32   s_wtime;                  /* Write time */
        __le16   s_mnt_count;              /* Mount count */
    [......]
```

The `__le32` data types indicate that the representation is in little-endian order. As evident from its definition in the kernel source, the Ext4 superblock defines a number of properties to characterize the filesystem. This contains information such as the total number of blocks and block groups in the filesystem, the total number of used and unused blocks, the block size, the total number of used and unused inodes, the filesystem state, and a lot more. The information contained in a superblock is of utmost importance as it is the first thing that is read when mounting a filesystem. Given its critical nature, multiple copies of the superblock are kept at different locations.

Most fields in the superblock definition are easily understood. Some interesting fields are explained here:

- **Block size calculation**: The block size of the Ext4 filesystem is computed using this 32-bit value. The block size is calculated as follows:

  *Ext4 block size = 2 ^ (10 + s_log_block_size)*

  The minimum block size of an Ext4 filesystem can be 1 KB when `s_log_block_size` is zero. The Ext4 filesystem supports a maximum block size of 64 KB.

- **Block clusters**: Even though the capacity of disk drives has grown exponentially in the last few years, the Ext4 filesystem works with block sizes of a few kilobytes. The bigger the drive, the larger the number of blocks and their overhead. As a workaround, the Ext4 developers added the functionality of block clusters in Ext4. Instead of allocating single 4 KB blocks, a filesystem can allocate blocks in larger groups using the concept of block groups. The Ext4 filesystem maintains the mapping between these larger blocks and 4 KB blocks. This feature is known as **bigalloc**. The block cluster size can be specified at filesystem creation time and is stored in `s_log_cluster_size`.

- **Filesystem state and checks**: The filesystem consistency check can be triggered in three cases. The `s_mnt_count` field indicates the number of times the filesystem has been mounted since the last consistency check was run. The `s_max_mnt_count` field imposes a hard limit on the number of mounts, beyond which a consistency check is mandatory. The filesystem state is saved in `s_state`. It can be one of the following:

  - `cleanly unmounted`

  - `errors detected`

  - `orphans being recovered`

  If the filesystem state in `s_state` is not clean, the check is enforced automatically. The date of the last consistency check is saved in `s_lastcheck`. If the time specified in the `s_checkinterval` field has passed since the last check, the consistency check is enforced on the filesystem.

- **Magic signatures**: Different filesystems use the concept of a magic number that appears at a certain offset. Different tools use this number as a way of identifying a particular filesystem type. The `s_magic` field in the superblock contains this magic number. For Ext4, its value is `0xEF53`. The `s_rev_level` and `s_minor_rev_level` fields are used to differentiate between filesystem versions.

- **Block reservation**: These are the default user and group IDs for reserved blocks. These default to a value of `0` (root user). An Ext4 filesystem reserves 5% of filesystem blocks for the super or root user. This is done so that the root user processes continue to run, even if the non-root processes cannot write to the filesystem.

- **First inode number**: This is the first inode number that can be used for regular files and directories. This value is usually `11`, which belongs to the `lost+found` directory on an Ext4 filesystem.

- **Filesystem UUID**: This is a 128-bit value that is used as a unique volume identifier for an Ext4 filesystem. On systems where drives are added and removed often, the device names (such as `sda` and `sdb`) can often change, resulting in confusion and incorrect mount points. The UUID is a unique identifier for a filesystem and can be used in `/etc/fstab` to mount filesystems.

- **Compatible features**: Both these values are 32-bit. The s_feature_compat field contains a 32-bit bitmask of compatible features. The filesystem is free to support the features defined in this field. On the other hand, if any feature defined in s_feature_incompat is not understood by the kernel, the filesystem mount operation will not succeed.

### Data block and inode bitmaps

The Ext4 filesystem uses a negligible amount of space to organize some internal structures. Most of the space in a filesystem is used for storing user data. The Ext4 filesystem stores user data in data blocks. As we learned in *Chapter 2*, the metadata of each file is stored in a separate structure, called an inode. The inodes are also stored on disk, albeit in a reserved space. Inodes are unique in a filesystem. Every filesystem uses a technique to keep track of allocated and available inodes. Similarly, there has to be a method through which the number of allocated and free blocks can be tracked.

Ext4 uses a bitmap as an allocation structure. A bitmap is a sequence of bits. Separate bitmaps are used to track the number of inodes and data blocks. The data block bitmap tracks the usage of data blocks within the block group. Similarly, the inode bitmap keeps track of entries in the inode table. A bit value of 0 indicates that the block or inode is available for use. A value of 1 indicates that the block or inode is occupied.

The bitmaps for both the inode and data block are of one block each. As a byte is composed of 8 bits, that means, for the default block size of 4 KB, the block bitmap can represent a maximum of 8 x 4 KB = 32,768 blocks per group. This can be verified in the output of mkfs or through the tune2fs program.

### Inode tables

In addition to inode bitmaps, a block group also contains an inode table. The inode table spans a series of consecutive blocks. The definition of an Ext4 inode is present in the fs/ext4/ext4.h file:

```
struct ext4_inode {
        __le16  i_mode;             /* File mode */
        __le16  i_uid;              /* Low 16 bits of Owner Uid */
        __le32  i_size_lo;          /* Size in bytes */
        __le32  i_atime;            /* Access time */
        __le32  i_ctime;            /* Inode Change time */
        __le32  i_mtime;            /* Modification time */
        __le32  i_dtime;            /* Deletion Time */
        __le16  i_gid;              /* Low 16 bits of Group Id */
        __le16  i_links_count;      /* Links count */
        __le32  i_blocks_lo;        /* Blocks count */
        __le32  i_flags;            /* File flags */
    [............]
```

The Ext4 inode has a size of 256 bytes. Some fields of particular interest are as follows:

- **Ownership**: The i_uid and i_gid fields serve as the user and group identifiers.

- **Timestamps**: The timestamps for a particular file are saved in i_atime, i_ctime, and i_mtime. These describe the last access time, inode change time, and data modification time, respectively. The file deletion time is saved in i_dtime. These 4 fields are 32-bit signed integers that represent the elapsed seconds since the Unix epoch time, January 1, 1970, 00:00:00 UTC. For calculating time with subsecond accuracy, the i_atime_extra, i_mtime_extra, and i_ctime_extra fields are used.

- **Hard links**: In contrast to a soft link, a hard link points to a file by its inode number. The hard link count of a file is defined in the i_links_count field. This is a 16-bit value, which means that Ext4 allows for a maximum of 65K hard links for a file.

- **Data block pointers**: In addition to some generic metadata, an inode also holds information about the on-disk locations of data blocks. This information is kept in i_block, which is an array of length EXT4_N_BLOCKS. The value of EXT4_N_BLOCKS is 15. As discussed in *Chapter 2*, an inode structure uses pointers for block addressing. First, 12 pointers point directly to block addresses and are called **direct pointers**. The next three pointers are indirect pointers. An **indirect pointer** points to a block of pointers. The 13th, 14th, and 15th pointers provide single-, double-, and triple-level indirection.

### Group descriptors

Group descriptors are stored just after the superblock in the filesystem layout. Every block group has a group descriptor associated with it, so there are as many group descriptors as the number of block groups. It's important to understand that the block group descriptors describe the contents of each block group in the filesystem. This means they include information about the local, as well as all the other block groups in the filesystem. The group descriptor structure is defined in fs/ext4/ext4.h:

```
struct ext4_group_desc
{
        __le32  bg_block_bitmap_lo;     /* Blocks bitmap block */
        __le32  bg_inode_bitmap_lo;     /* Inodes bitmap block */
        __le32  bg_inode_table_lo;      /* Inodes table block */
        __le16  bg_free_blocks_count_lo;/* Free blocks count */
        __le16  bg_free_inodes_count_lo;/* Free inodes count */
        __le16  bg_used_dirs_count_lo;  /* Directories count */
        __le16  bg_flags;               /* EXT4_BG_flags (INODE_
UNINIT, etc) */
[............]
```

Some important fields are described further:

- **Bitmap location**: The group descriptors contain information about the on-disk locations of block bitmaps, inode bitmaps, and the inode table. This information is stored in the following fields in the form of the least and most significant bits. The least significant bits are stored in bg_block_bitmap_lo, bg_inode_bitmap_lo, and bg_inode_table_lo, whereas the most significant bits are stored in bg_block_bitmap_hi, bg_inode_bitmap_hi, and bg_inode_table_hi.

- **Block and inode usage**: The group descriptors also contain information about the number of free blocks, inodes, and directories. This is also stored in the form of the least and most significant bits. The fields used to store this information are bg_free_blocks_count_lo, bg_free_blocks_count_hi, bg_free_inodes_count_lo, bg_free_inodes_count_hi, bg_used_dirs_count_lo, and bg_used_dirs_count_hi.

As each block group descriptor includes information about both local and non-local block groups, it contains a descriptor for each block group in the filesystem. Because of this, the following information can be determined from any single block group:

- The number of free blocks and inodes

- The location of the inode table in the filesystem

- The location of block and inode bitmaps

### Reserved GDT blocks

One of the most useful features of an Ext4 filesystem is its on-the-fly expansion. The size of an Ext4 filesystem can be increased on the fly without any disruption. The reserved **group descriptor table (GDT)** blocks are put aside at the time of filesystem creation. This is done to aid in the process of filesystem expansion. Increasing the size of the filesystem involves the addition of physical disk space and the creation of filesystem blocks in the newly added disk space. This also means that to accommodate the newly added space, more block groups and group descriptors will be required. These reserved GDT blocks are used when an Ext4 filesystem is to be extended.

### Journaling modes

Like most filesystems, Ext4 also implements the concept of journaling to prevent data corruption and inconsistencies in the case of a system crash. The default journal size is typically just a few megabytes. The journaling in Ext4 uses the generic journaling layer in the kernel, known as the **journaling block device (JBD** or **JBD2)**. If you've ever checked the top I/O consuming processes on a busy Linux system, you might have seen the jbd2 process in that list. This is the kernel thread responsible for updating the Ext4 journal.

Ext4 offers a great deal of flexibility when it comes to journaling. The Ext4 filesystem supports three journaling modes. Depending upon the requirements, the journaling mode can be changed if required. By default, journaling is enabled at the time of filesystem creation. If desired, it can be disabled later. The different journaling modes are explained here:

- **Ordered**: In ordered mode, only metadata is journaled. The actual data is directly written to disk. The order of the operations is strictly followed. First, the metadata is written to the journal; second, the actual data is written to disk; and last, the metadata is written to disk. If there is a crash, filesystem structures are preserved. However, the data being written at the time of the crash may be lost.

- **Writeback**: The writeback mode also only journals metadata. The difference is that actual data and metadata can be written in any order. This is a slightly more risky approach than ordered mode but offers much better performance.

- **Journal**: In journal mode, both data and metadata are written to the journal first, before being committed to the disk. This offers the highest level of security and consistency but can adversely affect performance, as all write operations have to be performed twice.

The default journaling mode is *ordered*. If you want to change the journal mode, you'll need to unmount the filesystem and add the desired mode in the corresponding `fstab` entry. For instance, to change the journaling mode to *writeback*, add `data=writeback` against the relevant filesystem entry in the `/etc/fstab` file. Once done, you can verify the journaling mode as follows:

```
[root@linuxbox ~]# mount |grep sdc
/dev/sdc on /mnt type ext4 (rw,relatime,data=writeback)
[root@linuxbox ~]#
```

You can also display information about the filesystem journal using the `logdump` command from `debugfs`. For instance, you can check the journal for the `sdc` device as follows:

```
[root@linuxbox ~]#  debugfs -R 'logdump -S' /dev/sdc
debugfs 1.44.6 (5-Mar-2019)
Journal features:          journal_64bit journal_checksum_v3
Journal size:              32M
Journal length:            8192
Journal sequence:          0x00000005
Journal start:             1
Journal checksum type:     crc32c
Journal checksum:          0xb78622f2
Journal starts at block 1, transaction 5
Found expected sequence 5, type 1 (descriptor block) at block 1
Found expected sequence 5, type 2 (commit block) at block 13
Found expected sequence 6, type 1 (descriptor block) at block 14
[.............]
```

## Filesystem extents

We've covered the use of indirect pointers to address large files. Through the use of indirect pointers, an inode can keep track of data blocks that contain the file contents. For large files, this approach becomes a bit inefficient. The higher the number of blocks occupied by a file, the higher the number of pointers required to keep track of each block. This creates a complex mapping scheme and increases the metadata usage per file. As a result, some operations on large files are performed rather slowly.

Ext4 makes use of extents to address this problem and reduce the metadata required to keep track of data blocks. An **extent** is a pointer plus a length of blocks – basically, a bunch of contiguous physical blocks. When using extents, we only need to know the address of the first and last block of this contiguous range. For instance, let's say that we're using an extent size of 4 MB. To store a 100 MB file, we can allocate 25 contiguous blocks. Since the blocks are contiguous, we only need to remember the address of the first and last blocks. Assuming a block size of 4 KB, while using pointers, we would need to create an indirect mapping of 25,600 blocks to store a 100 MB file.

## Block allocation policies

When it comes to filesystem performance, fragmentation is a silent killer. The Ext4 filesystem uses several techniques to improve the overall performance and reduce fragmentation. The block allocation policies in Ext4 ensure that related information exists within the same filesystem block group.

When a new file is to be created and saved, the filesystem will need to initialize an inode for that file. Ext4 will then select an appropriate block group for that file. The design of Ext4 makes sure that maximum effort is made to do the following:

- Allocate the inode in the block group that contains the parent directory of the file
- Allocate a file to the block group that contains the file's inode

Once a file has been saved on disk, after some time, the user wants to add new data to the file. Ext4 will start a search for free blocks, from the block that was most recently allocated to the file.

When writing data to an Ext3 filesystem, the block allocator only allocated a single 4 KB block at a time. Assuming a block size of 4KB, for a single 100 MB file, the block allocator would need to be called 25,600 times. Similarly, when a file is extended and new blocks are allocated from the block group, they can be in random order. This random allocation can result in excessive disk seeking. This approach does not scale well and causes fragmentation and performance issues. The Ext4 filesystem offers a significant improvement on this through the use of a multi-block allocator. When a new file is created, the multi-block allocator in Ext4 allocates multiple blocks in a single call. This reduces the overhead and increases performance. If the file uses those blocks, the data is written in a single multi-block extent. If the file does not use the extra allocated blocks, they are freed.

The Ext4 filesystem uses delayed allocation and does not allocate the blocks immediately upon a write operation. This is done because the kernel makes heavy use of the page cache. All operations are first performed in the kernel's page cache and then flushed to disk after some time. Using delayed allocation, the blocks are only allocated when data is actually being written to disk. This is extremely useful as the filesystem can then allocate contiguous extents for saving the file.

Ext4 tries to keep the data blocks of a file in the same block group as its inode. Similarly, all inodes in a directory are placed in the same block group as the directory.

### Examining the result of an mkfs operation

Let's summarize our discussion about Ext4 and see what happens when we create an Ext4 filesystem using mkfs. The following command was run on a disk of only 1 GB:

```
[root@linuxbox ~]# mkfs.ext4 -v /dev/sdb
mke2fs 1.44.6 (5-Mar-2019)
fs_types for mke2fs.conf resolution: 'ext4'
Discarding device blocks: done
Filesystem label=
OS type: Linux
Block size=4096 (log=2)
Fragment size=4096 (log=2)
Stride=0 blocks, Stripe width=0 blocks
65536 inodes, 262144 blocks
13107 blocks (5.00%) reserved for the super user
First data block=0
Maximum filesystem blocks=268435456
8 block groups
32768 blocks per group, 32768 fragments per group
8192 inodes per group
Filesystem UUID: ebcfa024-f87b-4c52-b3e1-25f1d4d31fec
Superblock backups stored on blocks:
        32768, 98304, 163840, 229376
Allocating group tables: done
Writing inode tables: done
Creating journal (8192 blocks): done
Writing superblocks and filesystem accounting information: done
```

Let's examine the output.

As the man page of mkfs.ext4 will tell you, the discarding device blocks feature is especially useful for solid-state drives. By default, the mkfs command will issue a TRIM command to inform the underlying drive to erase unused blocks.

The filesystem consists of 262,144 blocks of 4 KB each. The total number of inodes in the filesystem is 65,536. The UUID can be used in `fstab` to mount the filesystem.

The stride and stripe widths are used when the underlying storage is a RAID volume.

An Ext4 filesystem will by default reserve 5% space for the superuser.

We see that the filesystem has divided the 262,144 blocks into 8 block groups. There are a total of 32,768 blocks per group. Each block has 8,192 inodes. This is in line with the total number of inodes mentioned earlier – that is, 8 x 8,192 = 65,536.

The copies of the superblock structure are stored on multiple blocks. The filesystem will always be mounted using the primary superblock. But if the primary superblock gets corrupted for some reason, the filesystem can be mounted using the backups saved on different block locations. The filesystem journal occupies 8,192 blocks, which gives us a journal size of 8,192 x 4 KB = 32 MB.

The extended filesystem is one of the oldest Linux-specific software projects. Over the years, it has gone through several enhancements in terms of reliability, scalability, and performance. Most of the concepts associated with Ext4, such as journaling, the use of extents, and delayed allocation, also apply to XFS, although XFS uses different techniques to implement these features. Like all block-based filesystems, Ext4 divides the available disk space into fixed-size blocks. Being a native filesystem, it makes extensive use of the structures defined in VFS and implements them as per its own design. Because of its proven track record of stability, it is the most used filesystem across Linux distributions.

## Network filesystem

The evolution of computer networks and network protocols made remote file sharing possible. This gave rise to the concept of distributed computing and client-server architectures, which can be referred to as distributed filesystems. The idea was to store data on a central location on one or more servers. There are multiple clients that request access to this data through different programs and protocols. This includes protocols such as **File Transfer Protocol** (FTP) and **Secure File Transfer Protocol** (SFTP). The use of these programs makes it possible to transfer data between two machines.

As compared to any traditional filesystem, a filesystem that uses the distributed approach will require some additional elements for its functioning. We've seen that processes make use of the generic system call layer to issue read or write requests. In the case of conventional filesystems, both the process (which issues the request) and the storage (which serves that request) are part of the same system. In distributed systems, there can be a dedicated client-side application that is used for accessing the filesystem. In response to a generic system call such as `read ()`, the client side will send a message to the server requesting read access to a particular resource.

One of the oldest filesystems to follow this approach is the **Network Filesystem**, shortened as **NFS**. The NFS protocol was created by Sun Microsystems in 1984. NFS is a distributed filesystem, which allows accessing files stored in a remote location. NFS version 4 is the most recent version of the protocol. Since the communication between the client and server is over a network, any request by the clients will traverse all the layers in the **Open Systems Interconnection (OSI)** model.

## NFS architecture

From an architecture standpoint, NFS has three major components:

- **Remote Procedure Calls (RPCs)**: To allow processes to send and receive messages to and from each other, the kernel offers different **inter-process communication (IPC)** mechanisms. The NFS uses RPCs as a method of communication between an NFS client and server. RPC is an extension of the IPC mechanism. As the name suggests, in RPC, the procedure called by the client doesn't need to be in the same address space as the client. It can be in a remote address space. The RPC service is implemented at the session layer.

- **External Data Representation (XDR)**: NFS uses XDR as the standard for encoding binary data at the presentation layer of the OSI model. The use of XDR ensures that all stakeholders use the same language when communicating. The use of a standardized method for transferring data is necessary as data representation may differ between the two systems. For instance, it is possible that the NFS participants may be architecturally different and have different endianness. For instance, if data is being transmitted from a system using big-endian architecture to a system using little-endian architecture, the bytes will be received in the reverse order. XDR uses a canonical method of data representation. When an NFS client needs to write data on an NFS server, it will convert the local representation of relevant data into its equivalent XDR encoding. Similarly, when this XDR-encoded data is received by the server, it will decode and convert it back into its local representation.

- **NFS procedures**: All the NFS operations function at the application layer of the OSI model. The procedures defined at this layer specify the different tasks that can be performed on files residing on the NFS server. These procedures include file operations, directory operations, and filesystem operations.

*Figure 3.8* depicts the route taken by an I/O request when using NFS:

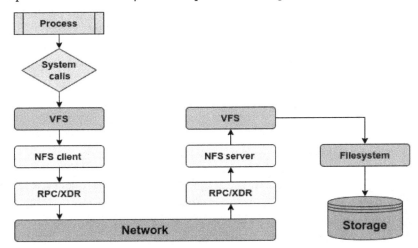

Figure 3.8 – Flow of an I/O request in NFS

In version 2, NFS used UDP as the underlying transport protocol and as a result, NFS v2 and v3 were stateless. An advantage of this approach was the slightly better performance because of the lower overhead when using UDP. Since version 4, the default protocol has been changed to TCP. Mounting NFS shares using TCP is a more reliable option. NFS version 4 is stateful, which means that both the NFS client and server maintain information about open files and file locks. In the case of a server crash, both the client and server side work to recover the state prior to the failure. NFS version 4 also introduced the **compound** request format. By using the compound request format, the NFS client can combine several operations into a single request. The compound procedure acts as a wrapper to coalesce one or more operations into a single RPC request.

Like any regular filesystem, the NFS also needs to be mounted to establish a logical connection between the client and server. This mount operation is a bit different from a local filesystem. While mounting an NFS filesystem, we do not need to create a filesystem since the filesystem already exists on the remote side. The mount command will include the name of the remote directory to be mounted. In NFS terms, this is called an **export**. The NFS server keeps a list of filesystems that can be exported and a list of hosts that are allowed to access these exports.

The NFS server uses a special structure to uniquely identify a file. This structure is known as a **file handle**. This handle makes use of an inode number, a filesystem identifier, and a generation number. The generation number plays a critical role in this identification process. Let's say that file A had an inode number of 100 and was deleted by the user. A new file, say B, was created and was assigned the recently freed inode number, 100. When trying to access a file using its file handle, this can cause confusion, as now, file B uses the inode number that was previously assigned to file A. Because of this, the file handle structure also uses a generation number. This generation number is incremented every time an inode is reused by the server.

## Comparing NFS with regular block filesystems

Network filesystems are also referred to as *file-level storage*. As such, I/O operations performed on an NFS are called file-level I/O operations. Unlike block filesystems, file-level I/O doesn't specify the block address of a file when requesting an operation. Keeping track of the exact location of the file on the disk is the job of the NFS server. Upon receiving the request from the NFS client, the NFS server will convert it into a block-level request and perform the requested operation. This does introduce additional overhead and is one of the major reasons that the performance of an NFS pales in comparison to a regular block filesystem. In the case of block filesystems and storage, applications have the freedom to decide how filesystem blocks will be accessed or modified. For NFS, the management of filesystem structures is entirely the responsibility of the NFS server.

We can see the differences between an NFS and a block filesystem:

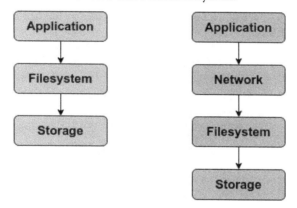

Figure 3.9 – NFS versus block filesystem

To sum up, NFS is one of the most popular protocols for remote file sharing. It is distributive in nature and follows a client-server architecture. The request from an NFS client ends up on the NFS server after traversing the entire network stack. To standardize data representation between the client and server, NFS uses XDR for data encoding at the presentation layer of the OSI model. Although it lags in performance when compared to regular block storage, it is still used in most enterprise infrastructures, mainly for backups and archiving.

# FUSE – a unique method for creating filesystems

We've talked about how the kernel divides the system into two portions: user space and kernel space. All the privileged system resources reside in kernel space. The kernel code, including that of filesystems, also exists in kernel space in a separate area of memory. It is not possible for normal user-space application programs to access it. The distinction between user space and kernel space programs restricts any regular process from modifying the kernel code.

Although this approach is essential to the kernel's design, it does create a few problems in the development process. Consider the example of any filesystem. As all the filesystem code exists in the kernel space, in the case of a bug in the filesystem code, it is extremely difficult to perform any troubleshooting or debugging because of this segregation. Any operations on the filesystem also need to be performed by the root user.

The **filesystem in user space** (**FUSE**) framework was designed to address some of these limitations. Through the use of the FUSE interface, filesystems can be created without tinkering with the kernel code. As such, the code for such filesystems only exists in user space. Both the actual data and metadata on the filesystem are managed by user-space processes. This is extremely flexible as it allows non-privileged users to mount the filesystem. It's important to note that FUSE-based filesystems can be stackable, meaning that they can be deployed on top of existing filesystems such as Ext4 and XFS. One of the most widely used FUSE-based solutions that makes use of this approach is **GlusterFS**. GlusterFS operates as a user-space filesystem and can be stacked on top of any existing block-based filesystem such as Ext4 or XFS.

The functionality provided by FUSE is achieved using a kernel module (`fuse.ko`) and a user-space daemon using the `libfuse` library. The FUSE kernel module is responsible for registering the filesystem with VFS. The interaction between the user-space daemon and the kernel is achieved using a character device, `/dev/fuse`. This device plays the role of a bridge between the user-space daemon and the kernel module. The user-space daemon will read from and write requests to this device:

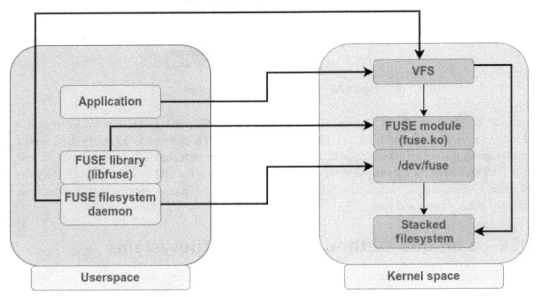

Figure 3.10 – The FUSE approach

When a process in user space performs any operation on a FUSE filesystem, the relevant system call is sent to the VFS layer. Upon checking that this corresponds to a FUSE-based filesystem, VFS will forward this request to the FUSE kernel module. The FUSE driver will create a request structure and put it in the FUSE queue in /dev/fuse. The communication between the kernel module and libfuse library is achieved using a special file descriptor. The user-space daemon will open the /dev/fuse device to process the result. If the FUSE filesystem is stacked on top of an existing filesystem, then the request will again be routed to the kernel space so that it can be passed to the filesystem underneath.

FUSE filesystems are not as robust as traditional filesystems but they offer a great deal of flexibility. They are easy to deploy and can be mounted by non-privileged users. Since the filesystem code is in user space, it is easier to troubleshoot and make changes. Even in the case of a bug in the code, the functionality of the kernel will not be impacted.

## Summary

Having covered the workings of VFS in the first two chapters, this chapter gave you an introduction to common filesystems and their concepts. The Linux kernel is capable of supporting around 50 filesystems and covering each one of them is an impossible task. We maintained our focus on the native filesystems in Linux, as the kernel is capable of supporting them out of the box. We explained some features that are common among a group of filesystems, such as journaling, CoW mechanisms, and FUSE. The major focus of this chapter was the working and internal design of the extended filesystem. The extended filesystem has been around since kernel version 0.96 and is the most widely deployed filesystem on computing platforms. We also shed some light on the architecture of network filesystems and explained the differences between file and block storage. At the end, we discussed FUSE, which offers an interface for user space programs to export a filesystem to the Linux kernel.

With this chapter, we've now completed our exploration of VFS and filesystem layers in the kernel. This brings the curtains down on *Part 1* of this book. I would like to think that this has been a good learning journey so far and I hope that it stays that way. The second part of this book, starting from *Chapter 4*, will focus on the block layer in the kernel, which provides an upstream interface to the filesystems.

# Part 2: Navigating Through the Block Layer

This part introduces the role of the block layer in the Linux kernel. The block layer is a key part of the kernel's storage stack because the interfaces implemented in the block layer are used by the user space applications to access the available storage devices. This part will explain the block layer and its major components, such as the device mapper framework, block devices, block layer data structures, the multi-queue framework, and the different I/O schedulers.

This part contains the following chapters:

# 4

# Understanding the Block Layer, Block Devices, and Data Structures

The first three chapters of this book were centered around the first component of the kernel's I/O hierarchy, which is the VFS layer. We explained the functions and purpose of VFS, as well as how it serves as an intermediary layer between the generic system call interface and filesystems, along with its primary data structures. In addition, we discussed the filesystems that can be found under the VFS layer and introduced some of the essential concepts associated with them.

We'll now turn our focus to the second major section in the kernel's storage hierarchy: the block layer. The block layer deals with block devices and is responsible for handling I/O operations performed on block devices. All the user-space programs use the block layer interfaces to address and access the underlying storage devices. Over the last decade or so, physical storage media has undergone a significant transformation, shifting from slower mechanical drives to faster flash drives. Consequently, the block layer within the kernel has undergone substantial modifications. As performance is a critical factor when it comes to storage hardware, several enhancements have been made to the kernel code to enable disk drives to realize their full potential. In this chapter, we're going to introduce the block layer, define block devices, and then dive into the major data structures in the block layer.

Here's a summary of what follows:

- Explaining the role of the block layer
- Defining block devices
- The defining characteristics of a block device
- Representation of block devices
- Looking at the major data structures in the block layer
- The journey of an I/O request in the block layer

# Technical requirements

The Linux kernel's block layer is a slightly complex topic. A good understanding of the material presented in the first three chapters will help you comprehend the interaction between the block layer and various filesystems. Having experience with the C programming language will help you understand the code presented in this chapter. Additionally, any practical experience with a Linux system will enhance your understanding of the concepts discussed herein.

If you want to download the kernel source, you can download it from `https://www.kernel.org`. The code segments referred to in this chapter and book are from kernel `5.19.9`.

# Explaining the role of the block layer

The block layer is tasked with implementing the kernel interfaces that enable filesystems to interact with storage devices. In the context of accessing physical storage, applications use block devices, and any requests to access data on these devices are managed by the block layer. The kernel also contains a mapping layer just above the block layer. This layer provides a flexible and powerful way to map one block device to another, allowing for operations such as creating snapshots, encrypting data, and creating logical volumes that span multiple physical devices. The interfaces that are implemented in the block layer are central to managing physical storage in Linux. The device files for block devices are created in the `/dev` directory.

Like VFS, abstraction is the core function of the block layer. The VFS layer allows applications to make generic requests for interacting with files without having to worry about the underlying filesystem. In a similar vein, the block layer allows applications to access storage devices uniformly. The choice of backend storage medium is not a point of concern for the application.

To highlight the major functions of the block layer, let's build on the storage hierarchy we defined when describing VFS. The following figure outlines the major components of the block layer:

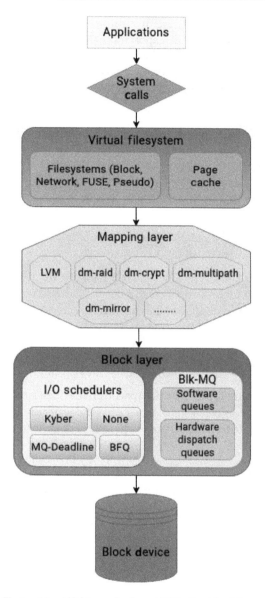

Figure 4.1 – I/O hierarchy from VFS to the block layer

Let's take a brief look at these functions:

- The block layer provides an upstream interface to filesystems and allows them to access a diverse range of storage devices uniformly. Likewise, it implements a downstream interface for drivers and storage devices by providing them with a single point of entry from all applications.

- As we'll see in this chapter, the block layer incorporates several tricky structures to provide its services generically. Probably the most important of them is the `bio` structure. The filesystem layer creates a bio structure to represent the I/O request and passes it down to the block layer. The bio structure is responsible for transporting all I/O requests to the driver. The mapping layer is in charge of providing an infrastructure to map physical block devices to logical devices. The mapping layer can achieve this by using the device mapper framework in the kernel. The device mapper lays the groundwork for several technologies in the kernel. These include volume management, multipathing, thin provisioning, encryption, and software RAID. The most well known of these is **logical volume management** (**LVM**). The device mapper creates every logical volume as a mapped device. LVM provides a great deal of flexibility to storage administrators and simplifies storage management.

- The `blk-mq` framework has become an important part of the block layer as it solved its performance limitations by isolating request queues for every CPU core. This framework is responsible for shepherding block I/O requests to multiple dispatch queues. We'll cover the `blk-mq` framework in more detail in *Chapter 5*.

- The block layer also includes several schedulers for handling I/O requests. These schedulers are pluggable and can be set for individual block devices. Non-multi-queue schedulers have been deprecated and are no longer supported in modern kernels. As we'll see in *Chapter 6*, these schedulers make use of several techniques to make intelligent decisions regarding I/O scheduling.

- Additionally, the block layer implements functions such as error handling and collecting I/O statistics for block devices.

At the heart of the block layer are block devices. Apart from devices that stream data, such as tape drives, most storage devices, such as mechanical drives, and solid-state drives flashcards, are considered block-based devices. Let's take a look at the defining traits of a block device and how they're represented in Linux.

# Defining block devices

There are two major ways that the kernel exchanges data with external devices. One method is to exchange one character at a time with the device. Devices addressed through such methods are known as character devices. Character devices are addressed using a stream of sequential data. They can be accessed by programs to perform input and output operations one character at a time. Due to the absence of random access methods, managing character devices is simpler for the kernel. Devices such as keyboards, text-based consoles, and serial ports are all examples of character devices.

Communicating through one character at a time is acceptable when the volume of data is low, such as when using serial ports or keyboards. A keyboard can only accept one character at a time, so the use of a character interface makes sense. But this approach becomes untenable when transferring large volumes of data. When writing to physical disks, we expect them to be able to address more than one character at a time and allow for data to be accessed randomly. The kernel addresses physical drives in fixed-size chunks, known as blocks. In addition to traditional disks, devices such as CD-ROMs and flash drives also use this approach. Such devices are known as block devices. Block devices are more complex to manage compared to character devices and require more careful considerations from the kernel. The kernel has to make critical decisions regarding the addressing and organization of block devices as these can significantly impact not only the block device but also the overall system performance.

Block devices can exist in memory. This can be achieved by creating a *ramdisk*. One of the most notable use cases of ramdisks is during the boot-up sequence in Linux systems. An **initial ramdisk** (**initrd**) is responsible for loading a temporary root filesystem in memory to aid in the boot process. A filesystem can be created on a ramdisk and mounted like any regular filesystem. The speed of RAM makes ramdisks incredibly fast. But due to the volatile nature of RAM, any data written to a ramdisk is maintained only until the device is powered on.

Although ramdisks are also block-based, they're rarely used. As you'll see, throughout this book, a block device is typically regarded as a persistent data storage medium with a filesystem layer on top.

All operations on block devices are performed by the kernel in fixed-size chunks of $N$ bytes, known as *blocks*, which is the currency of exchange when dealing with block devices. The actual value of $N$ varies across the stack since different layers in the kernel's I/O hierarchy use chunks of different sizes to address block devices. Because of this, the term block is defined in several ways, depending on its presence in the stack:

- **User-space applications**: As applications interact with the kernel space through standard system calls, the term **block** in this context refers to the amount of data read from and written to via system calls. Depending on the application, this can vary in size.

- **Page cache**: The kernel makes extensive use of the VFS page cache to improve the performance of read and write operations. Here, the fundamental unit of data transfer is a *page*, which is 4 KB in size.

- **Disk-based filesystems**: As explained in *Chapter 3*, a block represents the fixed number of bytes in which I/O operations are performed by the filesystem. Although filesystems allow for higher block sizes, often up to 64 KB, the block size for a filesystem is usually between 512 bytes and 4 KB because of the page size.

- **Physical storage**: On physical disks, the smallest addressable unit is known as a sector, which is usually 512 bytes. This sector is often further classified as logical or physical.

We discussed filesystem blocks in *Chapter 3*. Don't get confused; the filesystem block's size is not the fundamental unit of block I/O. The basic unit of block I/O is a sector. The data structures in the block layer define a variable of the `sector_t` type in the kernel code that represents an offset or size that is a multiple of 512. The `sector_t` variable is defined as an unsigned integer type that is large enough to represent the maximum number of sectors that can be addressed by the block device. It is used extensively throughout the block layer in structures such as "bio" to represent disk addresses and offsets.

To summarize, devices that are organized and addressed in terms of blocks are known as block devices. They allow for random access and offer superior performance compared to character devices. To take full advantage of block devices, the kernel has to make informed decisions regarding their addressing and organization.

Let's briefly go over some key features that define a block device.

# The defining characteristics of a block device

As discussed earlier, block devices allow far more advanced ways to handle I/O requests. Some of the defining characteristics of block devices are as follows:

- **Random access**: Block devices allow for random access. This means that the device can *seek* from one position to another.

- **Block size**: Block devices address and transfer data in fixed-sized blocks.

- **Stackability**: Block devices can be stacked through the use of the device mapper framework. This extends the basic functionalities of physical disks and allows for scaling logical volumes.

- **Buffered I/O**: Block devices use buffered I/O, which means that data is written to a buffer in memory before being written to the device. Read and write operations on block devices make extensive use of the page cache. Data that's read from the block device is loaded and kept in memory for a certain period. Similarly, any data to be written to a block device is written to the cache first.

- **Filesystems/partitioning**: Block devices can be partitioned into smaller logical units, with separate filesystems created on top of them.

- **Request queues**: Block devices implement the concept of request queues, which are responsible for managing the I/O requests that are submitted to a block device.

Let's see how block devices are represented in Linux.

# Looking at the representation of block devices

When discussing VFS, we saw that abstractions are at the heart of the kernel's I/O stack. The block layer is no exception to this rule. Regardless of the differences in the physical make and model, the kernel should be able to work with storage devices uniformly. To implement a standard interface for all devices, the operations should be independent of the properties of the underlying storage device.

As explained in *Chapter 1*, almost everything is represented in the form of a file, including hardware devices. A block device is a special file and is named as such because the kernel interacts with it using a fixed number of bytes. Depending on the nature of the devices, the files representing them are created and stored at specific locations in the system. The block devices in the system are present in the /dev directory. Filenames representing disk drives in the system start with sd, followed by a letter representing the order of discovery. The first drive is named sda and so on. Similarly, the first partition on the sda drive is represented as sda1. If we look at the sd* devices in /dev, notice that the file type is b, for block devices. You can also list block devices using the lsblk command, as shown in the following figure:

```
root@linuxbox:~# ls -l /dev/sd*
brw-rw---- 1 root disk 8,  0 Nov 28 18:37 /dev/sda
brw-rw---- 1 root disk 8,  1 Nov 28 18:37 /dev/sda1
brw-rw---- 1 root disk 8,  2 Nov 28 18:37 /dev/sda2
root@linuxbox:~#
```

Figure 4.2 – Major and minor numbers

Just before the modification timestamp, note the two numbers separated by a comma. The kernel represents block devices as a pair of numbers. These numbers are called major and minor numbers for a device. The *major number* identifies the driver associated with the device, whereas the *minor number* is used for differentiating between individual devices.

In the preceding figure, all three devices – sda, sda1, and sda2 – use the same driver and hence have the same major number, 8. The minor numbers – 0, 1, and 2 – are used to identify the driver instance for each device.

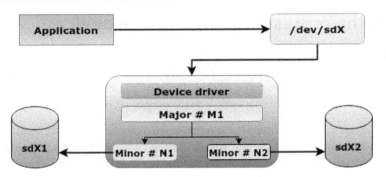

Figure 4.3 – Device major and minor numbers

The device files present in the /dev directory are hooked up to the corresponding device drivers to establish a communication link with the actual hardware. When a program interacts with the block device file, the kernel uses the major number to identify the appropriate driver for that device and sends the request. Since a driver can be responsible for handling multiple devices, there has to be a way through which the kernel can distinguish devices using the same major number. For this purpose, minor numbers are used.

We will now explore the primary data structures that are used in the block layer.

## Looking at data structures in the block layer

Dealing with block devices is fairly complex as the kernel has to implement features such as queue management, scheduling, and the ability to access data randomly. The speed of block devices is much higher than character devices. This makes block devices extremely performance-sensitive and the kernel has to make intelligent decisions to extract their maximum performance. Thus, it makes sense to handle the two devices differently. Because of this, there is an entire kernel subsystem dedicated to managing block devices. All this makes the block layer the most sophisticated piece of code in the Linux kernel.

Throughout this book, we've referred to the relevant pieces of kernel code so that you can familiarize yourself with the implementations of certain concepts. If you're interested in pursuing kernel development, this may seem like a good starting point. However, if you are more focused on theoretical understanding, the use of code might be mildly confusing. But it is essential to get a basic idea of how certain things are represented in the kernel. Talking specifically about the block layer, it is not possible to discuss all the structures that make up its intricate design. Nevertheless, we must highlight some more important constructs that give us a fair understanding of the representation and organization of block devices in the kernel.

Some of the major data structures that are used to work with block devices are as follows:

- `register_blkdev`
- `block_device`

- `gendisk`
- `buffer_head`
- `bio`
- `bio_vec`
- `request`
- `request_queue`

Let's take a look at them one by one.

## The register_blkdev function (block device registration)

To make block devices available for use, they must first be registered with the kernel. The registration process is performed by the `register_blkdev()` function, which is defined in `include/linux/blkdev.h`:

```
int __register_blkdev(unsigned int major, const char *name,
                void (*probe)(dev_t devt))
```

The `register_blkdev` function is used by block device drivers to register themselves and is a macro that's directed to `__register_blkdev`. The `__register_blkdev` function performs the actual registration process. The purpose of having a separate internal function is to provide additional error handling and validation before modifying the kernel's data structures.

The registration function performs the following tasks:

- It requests a major number from the kernel's dynamic major number allocation pool. The major number uniquely identifies the block device driver within the system.
- Once a major number has been successfully obtained, the function creates a `block_device` struct, which represents the block device driver. This struct contains information such as the major number, the name of the driver, and function pointers to various driver operations.

In summary, the `register_blkdev` function acts as a friendly interface through which block device drivers can initiate their registration processes with the kernel's block layer. It handles the necessary steps for acquiring a major number, creating a `block_device` struct, and establishing the necessary connections with the block layer.

## The block_device structure (representing block devices)

The block device is defined in `include/linux/blk_types.h` by the `block_device` structure in the kernel:

```
struct block_device {
        sector_t                bd_start_sect;
        sector_t                bd_nr_sectors;
        struct disk_stats __percpu *bd_stats;
        unsigned long           bd_stamp;
        bool                    bd_read_only;
        dev_t                   bd_dev;
        atomic_t                bd_openers;
        struct inode *          bd_inode;
[........]
```

The `block_device` structure instance is created when the device file is opened. A block device can be a whole disk or a single partition and the `block_device` structure can represent either. When using partitions, the individual partitions are identified through the `bd_partno` field. Since access to block devices happens through the VFS layer, the corresponding device files are also assigned an inode number. The inodes for block devices are virtual and are stored in the `bdev` virtual filesystem. The inode for a block device also contains information about its major and minor numbers.

The `block_device` structure also provides information about the device, such as its name, size, and block size. It also contains a pointer to the `gendisk` structure, which represents the disk, and a list of `request_queues` structures for handling I/O requests.

## The gendisk structure (representing physical disks)

An important field in the definition of the `block_device` structure is the `bd_disk` pointer, which points to the `gendisk` structure. The `gendisk` structure, which is defined in `include/linux/blkdev.h`, represents information about the disk and is used to implement the notion of a physical hard disk in the kernel.

The `gendisk` structure represents the disk's properties and the methods used to access it. It is used to register a block device and its associated I/O operations with the kernel, allowing it to communicate with the device:

```
struct gendisk {
        int major;
        int first_minor;
        int minors;
        char disk_name[DISK_NAME_LEN];
```

```
        unsigned short events;
        unsigned short event_flags;
[......]
```

gendisk can be considered a link between the block and filesystem interfaces mentioned previously and the hardware interface. There will be a block_device structure for representing the entire physical disk defined in gendisk. Similarly, there will be separate block_device structures that describe individual partitions within gendisk. Note that gendisk is allocated and controlled by the block device driver and registered with the kernel using the register_blkdev function. Once registered, the block device driver can use the gendisk structure to perform I/O operations on the device.

Let's look at some of the important fields of this structure:

- major: This field specifies the major number associated with the gendisk structure. As discussed earlier, the major number is used by the kernel to identify the driver responsible for handling the block device.

- first_minor: This field refers to the smallest minor number that is allocated to a given block device. This can be thought of as an offset from which the minor numbers for the different partitions of the device are allocated.

- minors: This field specifies the total number of minor numbers associated with the gendisk structure.

- fops: This field points to a structure of file operations that are associated with the gendisk structure. These file operations are used by the kernel to handle read, write, and other file operations on the device.

- private_data: This field is used by the driver to store any private data associated with the gendisk structure, such as any driver-specific information.

- queue: This field points to the request queue associated with the gendisk structure. The request queue is responsible for managing the I/O requests that are issued to the device. This makes it a very important field as it enables the kernel to associate a specific I/O queue with each block device. By having separate I/O queues for each block device, the kernel can manage multiple block devices independently and handle their I/O operations more efficiently. This allows the kernel to optimize performance, apply appropriate scheduling policies, and prevent I/O bottlenecks.

- disk_name: This field is a string that specifies the name of the device. The name is used by the kernel to identify the device and is usually displayed in system logs.

Let us move on to the next structure.

## The buffer_head structure (representing blocks in memory)

One of the defining features of a block device is its extensive use of the page cache. The read and write operations on a block device are performed in the cache. When an application reads from a block for the first time, the block is fetched from the physical disk into memory. Similarly, when a program wants to write some data, the write operation is first performed in the cache. It is written to the physical disk at a later stage.

The block that's read from the disk or to be written to the disk is stored in a buffer. This buffer is represented by the `buffer_head` structure, which is defined in `include/linux/buffer_head.h` in the kernel. We can say that this buffer is an in-memory representation of an individual block:

```
struct buffer_head {
        unsigned long b_state;
        struct buffer_head *b_this_page;
        struct page *b_page;
        sector_t b_blocknr;
        size_t b_size;
[..........]
```

The fields in the `buffer_head` structure contain the information required to uniquely identify a particular block in the block device. The fields in the `buffer_head` structure are described as follows:

- `b_data`: This field points to the start of the data buffer associated with `buffer_head`. The size of the buffer is determined by the block size of the filesystem.

- `b_size`: This field specifies the size of the buffer in bytes.

- `b_page`: This is a pointer to the page in memory where the block is stored. This field is typically used in conjunction with fields such as `b_data` and `b_size`, to manipulate the buffer data.

- `b_blocknr`: This field specifies the logical block number of the buffer in the filesystem. Each block in the filesystem is assigned a unique number called a logical block number. This number represents the order of the block in the filesystem, starting from 0 for the first block.

- `b_state`: This field is a bitfield that represents the state of the buffer. It can have several values. For instance, a value of `BH_Uptodate` indicates that the buffer contains up-to-date data, while a value of `BH_Dirty` indicates that the buffer contains dirty (modified) data and needs to be written to disk.

- `b_count`: This field keeps track of the number of users of `buffer_head`.

- `b_page`: This field points to the page in the page cache that contains the data associated with the `buffer_head` structure.

- `b_assoc_map`: This field is used by some filesystems to track which blocks are currently associated with `buffer_head`.

- b_private: This field is a pointer to private data associated with buffer_head. This can be used by the filesystem to store information related to the buffer.

- b_bdev: This field is a pointer to the block device that the buffer belongs to.

- b_end_io: This field is a function pointer that specifies the completion function for an I/O operation on a buffer, and is used to perform any necessary cleanup operations.

By default, as the filesystem's block size is equal to the page size, a single page in memory can hold a single block. If the block size is less than the page size, the page can hold more than one block.

The buffer head maintains a mapping between a page in memory and its corresponding on-disk version. Although it still holds important information, it was an even more integral component of the kernel before version 2.6. Back then, in addition to maintaining page-to-disk block mapping, it also served as a container for all I/O operations in the block layer. The use of buffer heads as an I/O container resulted in a significant amount of memory usage. When dealing with a large block of I/O requests, the kernel had to break it into smaller requests, each of which, in turn, had a buffer_head structure associated with them.

## The bio structure (representing active block I/Os)

Due to the limitations in the buffer_head structure, the bio structure was created to represent an ongoing block I/O operation. The bio structure has been the fundamental unit of an I/O in the block layer since kernel 2.5. When an application issues an I/O request, the underlying filesystem translates it into one or more bio structures, which are sent down to the block layer. The block layer then uses these bio structures to issue I/O requests to the underlying block device. The bio structure is defined in include/linux/blk_types.h:

```
struct bio {
        struct bio              *bi_next;
        struct block_device     *bi_bdev;
        unsigned int            bi_opf;
......
        unsigned short          bi_max_vecs;
        atomic_t                __bi_cnt;
        struct bio_vec          *bi_io_vec;
[..........]
```

Some particularly interesting fields are as follows:

- bi_next: This is a pointer to the next bio structure in a list and is used to link multiple bio structures that represent a single I/O operation. This is important to understand because a single I/O operation may need to be split into multiple bio structures.

- `bi_vcnt`: This field specifies the number of `bio_vec` structures that are being used to describe the I/O operation. Each `bio_vec` structure in the vector describes a contiguous block of memory that is transferred between the block device and the user space program.

- `bi_io_vec`: This is a pointer to an array of `bio_vec` structures that describes the location and length of the data buffers associated with the I/O operation. This lays the ground for performing `scatter-gather` I/O – that is, the data can be spread across multiple non-contiguous memory locations.

- `bi_vcnt`: This field specifies the number of data buffers associated with the I/O operation. Each data buffer is represented by a `bio_vec` structure, which contains a pointer to the memory buffer and the length of the buffer.

- `bi_end_io`: This is a pointer to a function that is called when the I/O operation completes. This function is responsible for cleaning up any resources associated with the I/O operation and waking up any processes that are waiting for the operation to complete.

- `bi_private`: This is a pointer to any private data associated with the I/O operation.

- `bi_opf`: This is a bit mask that specifies any additional options or flags associated with the I/O operation. This can include options such as *force synchronous I/O* or *disable write caching*.

When an I/O request is initiated by a user-space application, the bio structure keeps track of all the active I/O transactions at the block layer. Once the bio structure has been constructed, it is handed over to the block I/O layer through the `submit_bio` function. The `submit_bio()` function is used to submit I/O requests to block devices. Once the I/O has been submitted to the block device, it is added to a request queue. The `submit_bio()` function will not wait for the I/O to be completed.

It can be said that the bio structure acts as a bridge between the filesystem and the block device layer, enabling the filesystem to perform I/O operations on the block device.

## The bio_vec structure (representing vector I/O)

The `bio_vec` structure defines vector or scatter-gather I/O operations in the block layer.

The `bio_vec` structure is defined in `include/linux/bvec.h`:

```
struct bio_vec {
        struct page     *bv_page;
        unsigned int    bv_len;
        unsigned int    bv_offset;
};
```

The fields are described as follows:

- `bv_page`: This field holds a reference to the page structure (struct page) that contains the data to be transferred. As we explained in *Chapter 2*, a page is a fixed-size block of memory.

- `bv_offset`: This field holds the offset within the page where the data to be transferred starts.

- `bv_len`: This field holds the length of the data to be transferred.

**Scatter-gather I/O** involves transferring data between a device and the memory. Usually, data is read from or written to a single contiguous memory buffer. With scatter-gather I/O, the data is divided into smaller segments and spread across multiple non-contiguous memory buffers, known as **scatter lists**, for input operations. For output operations, the data is then gathered from these multiple non-contiguous memory buffers. The `bio_vec` structure is used to represent a scatter-gather I/O operation. The block layer may construct a single bio with multiple `bio_vec` structures, each representing a different physical page in memory and a different offset within that page.

## Requests and request queues (representing pending I/O requests)

When an I/O request is submitted to the block layer, the block layer creates a `request` structure to represent the request.

The `request` and `request_queue` structures are defined in `include/linux/blk-mq.h` and `include/linux/blkdev.h`, respectively:

```
struct request {
        struct request_queue *q;
        struct blk_mq_ctx *mq_ctx;
        struct blk_mq_hw_ctx *mq_hctx;
[.......]
```

Some of the major fields are explained here:

- `struct request_queue *q`: Each I/O request is added to the request queue of a block device. The q field here points to this request queue.

- `struct blk_mq_ctx *mq_ctx`: The `blk_mq_ctx *mq_ctx` field points to the software staging queues; this structure is allocated on a per-CPU core basis. Each CPU has a `blk_mq_ctx`, which is used to track the state of requests that are processed on that CPU.

- `struct blk_mq_hw_ctx *mq_hctx`: This field represents the hardware context with which a request queue is associated. This is used to keep track of which hardware queue the request belongs to.

- `struct list_head queuelist`: This is a linked list of requests that are waiting to be processed. When a request is submitted to the block layer, it is added to this list.

- `struct request *rq_next`: This is a pointer to the next request in the queue. It is used to link requests within the request queue.

- `sector_t sector`: This field specifies the starting sector number of the I/O operation.

- `struct bio *bio`: This field points to a `bio` structure that contains information about the I/O operation, such as its type (read or write).

- `struct bio *biotail`: This field points to the last `bio` structure in the queue. When a new bio is added to the queue, it is linked to the end of the list pointed to by `biotail`.

The `request_queue` structure represents the request queue associated with a block device. The request queue is responsible for managing all I/O requests that are issued to the block device:

```
struct request_queue {
        struct request         *last_merge;
        struct elevator_queue  *elevator;
        struct percpu_ref      q_usage_counter;
[..........]
```

Let's look at some of the important fields:

- `struct request *last_merge`: This field is used by the I/O scheduler to track the last request that was merged with another request.

- `struct elevator_queue *elevator`: This field points to the I/O scheduler for the request queue. The I/O scheduler determines the order in which requests are serviced.

- `struct percpu_ref q_usage_counter`: This field represents the usage counter for the request queue. The kernel uses a per-CPU counter to track the reference count of a resource on a per-CPU basis.

- `struct rq_qos *rq_qos`: This field points to a request queue of quality-of-service agreements that the request queue provides to the block device. These are used to prioritize I/O requests based on different criteria, such as the priority of the request.

- `const struct blk_mq_ops *mq_ops`: This structure contains function pointers that define the behavior of the request queue for multi-queue I/O schedulers.

- `struct gendisk *disk`: This field points to the `gendisk` structure associated with the request queue. `gendisk` represents a generic disk device.

Phew! There are too many of them. Let's summarize the role of each structure and see how they all work together.

# The journey of an I/O request in the block layer

The following table provides a concise overview of the structures that were covered in the previous section:

| Structure | Representation of | Description |
|---|---|---|
| `gendisk` | Physical disk | This is used to represent the physical device as a whole and contains information such as the disk's capacity and geometry |
| `block_device` | Block device | This represents a specific instance of a device and contains information such as the major and minor numbers, partitions, and the queue to handle I/O requests |
| `buffer_head` | Block of data in memory | This is used to track the data that is read from or written to a block device in memory |
| `request` | I/O request | This includes information such as the type of I/O operation and the starting block number |
| `request_queue` | Queue of I/O requests | This contains information about the current state of the queue, such as the number of requests waiting to be processed |
| `bio` | Block I/O | This is a higher-level I/O request and can include multiple request structures |
| `bio_vec` | Scatter-gather list of memory buffers | This is used as a part of the bio structure and describes an individual data buffer |

Table 4.1 – Summary of major block layer structures

Let's take a look at the relationship between these structures when a process issues an I/O request:

1. When an application writes data in a buffer in its address space, the block layer creates a `buffer_head` structure to represent this data.

2. The block layer constructs a `bio` structure to represent the block I/O request and maps the `buffer_head` structure to the `bio_vec` structure. For each `bio`, the block layer creates one or more `bio_vec` structures to represent the data being read from or written to.

3. The `bio` structure is then added to the `request_queue` structure for the intended block device through the `request` structure.

4. The device driver for that device, which is registered through `register_blkdev`, dequeues the `bio` structure and schedules it for processing.

5. `bio` is then split into one or more `request` structures based on the block size of the device.

6. Each `request` object is then added to the `request_queue` structure of the corresponding device driver for processing.

7. After processing the request, the device driver writes the data to the physical storage.

8. Once the I/O request has been completed, the device driver notifies the block layer.

9. The block layer then updates the buffer cache and the associated data structures. It marks the request structure as completed and notifies any waiting processes that the I/O operation has been completed.

10. The corresponding `buffer_head` structures are updated to reflect the current state of the data on the block device.

The block layer, with its intricate design, makes use of some complex structures to work with block devices. We covered some major structures to help you understand how things work under the hood. Each structure defines a ton of fields in its definition; we've tried to highlight some to get a gist of things.

It's important to note that the request queues in older kernels were single-threaded and were not able to exploit the capabilities of modern hardware. The Linux kernel added multi-queue support in version `3.13`. The framework for implementing multi-queue support is known as `blk-mq`. We're going to cover the multi-queue framework in the next chapter.

## Summary

The first part of this book, which included *Chapters 1*, *2*, and *3*, dealt with VFS and filesystems. The second part of this book, which constitutes *Chapters 4*, *5*, and *6*, is all about the block layer. This chapter introduced the role of the block layer in the Linux kernel. The **block layer** is the kernel subsystem and is in charge of managing I/O operations performed on **block** devices. The kernel's block device interface is central to managing persistent storage on Linux. The user-space applications can access the block devices through block special devices in the /dev directory. Working with block devices is far more complicated than working with character devices, which can only work sequentially. Character devices have a single current position. Managing block devices is a far more complex task for the kernel as block devices must be able to move to any position to provide random access to data. Because of this, performance is a major concern when working with block devices. The Linux kernel provides a complex ecosystem of structures in the block layer for working with block devices.

In the next chapter, we will build on our understanding and see how an I/O request is served in the block layer. We'll also cover the device mapper and multi-queue frameworks in the kernel.

# 5

# Understanding the Block Layer, Multi-Queue, and Device Mapper

*"I feel the need... the need for speed." – Maverick in Top Gun*

*Chapter 4* introduced us to the role of the block layer in the kernel. We were able to see what constitutes a block device and explored the major data structures in the block layer. This chapter will build on that knowledge as we continue understanding the block layer.

This chapter will introduce you to two major concepts: the multi-queue block I/O mechanism and the device mapper framework. The kernel's block layer has undergone significant changes in recent years to tackle performance concerns. The introduction of the multi-queue framework was a significant milestone in this direction, as discussed in *Chapter 4*. Performance is a critical consideration when dealing with block devices, and the kernel has implemented various improvements to optimize disk drive performance. In *Chapter 4*, we looked at the request and response queue structures in the block layer, which handle the I/O requests for a block device. In this chapter, we'll start by introducing the single-request queue model, its performance limitations, and the challenges faced by the block layer when working with modern high-performing storage drives such as NVMe and SSDs. We'll also explain how the single-request queue model impacts the performance of multicore systems.

The second major topic of this chapter will be the mapping framework in the kernel, known as the device mapper. The device mapper framework in the kernel works in conjunction with the block layer and is responsible for mapping physical block devices to logical block devices. As we will see, the device mapper framework serves as the foundation for implementing various technologies, such as logical volume management, RAID, encryption, and thin provisioning. In the end, we'll also briefly discuss caching mechanisms in the block layer.

We will discuss the following main topics:

- The problem with single-request queues
- The multi-queue block I/O mechanism
- The device mapper framework
- Multi-tier caching in the block layer

# Technical requirements

In addition to the Linux operating system concepts we covered previously, the topics discussed in this chapter require a basic understanding of modern processors and storage technologies. Any practical experience in Linux storage administration will greatly enhance your understanding of certain aspects.

The commands and examples presented in this chapter are distribution-agnostic and can be run on any Linux operating system, such as Debian, Ubuntu, Red Hat, Fedora, and others. There are quite a few references to the kernel source code. If you want to download the kernel source, you can download it from `https://www.kernel.org`. The code segments referred to in this chapter and book are from kernel `5.19.9`.

# Looking at problems with single-request queues

The operating system must handle block devices so that they operate at their full potential. An application may need to perform I/O operations on arbitrary locations on a block device, which requires seeking multiple disk locations and can prolong the operation's duration. When rotating mechanical drives, constant random accesses can not only degrade performance but also produce noticeable noise. Although still used in the modern day, interfaces such as **Serial Advanced Technology Attachment** (**SATA**) were the protocol of choice for mechanical drives. The original design of the kernel's block layer was meant for a time when mechanical drives were the medium of choice. These legacy hard drives could only handle a few hundred IOPs. Two things changed this: the ascendance of multi-core processors and the advancement in drive technologies. With these changes, the bottleneck in the storage stack shifted from the physical hardware to the software layers in the kernel.

In the legacy design, the kernel's block layer handled I/O requests in one of the following ways:

- The block layer maintained a single-request queue, a linked list structure, to handle I/O requests. New requests were inserted at the tail end of the queue. The block layer implemented techniques such as merging and coalescing (which we'll explain in the next chapter) on these requests before handing them over to the driver.
- In some cases, the I/O requests had to bypass the request queues and land directly on the device driver. This meant that all the processing done in the request queue would be performed by the driver. This usually resulted in a negative performance impact.

Even with the use of modern solid-state drives, this design suffered from major limitations. This approach further results in a three-fold problem:

- The request queue containing I/O requests didn't scale to handle the needs of modern processors. On multi-core systems, a single-request queue had to be shared between multiple cores. Therefore, to access the request queue, a locking mechanism was used. This global lock was used to synchronize shared access to the block layer request queue. To implement the different I/O handling techniques, a CPU core needed to acquire a lock to the request queue. This meant that if another core needed to operate on the request queue, it had to wait a considerable amount of time. All CPU cores remain in a state of contention for the request queue lock. It's not too difficult to see that this design made the request queue the single point of contention on multi-core systems.

- A single-request queue also introduces cache coherency problems. Each CPU core has its own L1/L2 cache, which may contain a copy of the shared data. When a CPU core modifies some data after acquiring a global lock to the request queue and updates said data in its cache, the other cores may still contain stale copies of the same data in their caches. As a result, modifications made by one core may not be promptly propagated to the caches of other cores. This leads to an inconsistent view of the shared data across different cores. When the global lock to the request queue is freed by a core, its ownership is transferred to another core already waiting for the lock. Although several cache coherency protocols exist, which ensure that caches maintain a consistent view of the shared data, the bottom line is that the single-queue design does not inherently provide mechanisms to synchronize the caches of different CPU cores. This increases the overall workload required to ensure cache coherency.

- This frequent switching of request queue locks between cores results in an increased number of interrupts.

All in all, the use of multiple cores meant that multiple execution threads would be simultaneously competing for the same shared lock. The higher the number of CPUs/cores in the system, the higher the lock contention for the request queue. A significant number of CPU cycles are wasted due to the spinning and contention involved in acquiring this lock. On multi-socket systems, this greatly reduces the number of IOPs.

*Figure 5.1* highlights the limitations of using the single queue model:

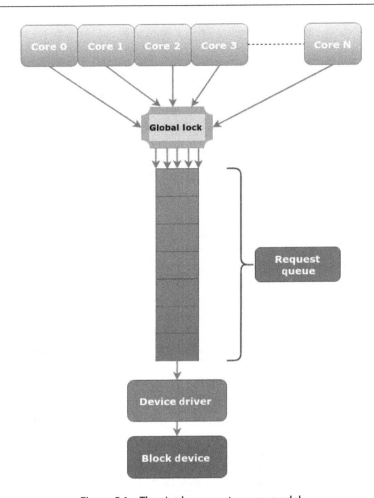

Figure 5.1 – The single-request queue model

From *Figure 5.1*, it becomes abundantly clear that regardless of the CPU core count and the type of underlying physical storage, the single queue block layer's design could not scale up to match their performance requirements.

In the past decade or so, enterprise storage environments have shifted to solid-state drives and non-volatile memory. These devices do not have mechanical parts and are capable of handling I/O requests in parallel. The design of these devices ensures that no performance penalty is observed when doing random access. With the emergence of flash drives as the preferred persistent storage medium, the traditional techniques that were used in the block layer for working with HDDs became obsolete. To fully leverage the enhanced capabilities of SSDs, the design of the block layer needed to mature accordingly.

In the next section, we'll see how the block layer has evolved to meet this challenge.

# Understanding the multi-queue block I/O framework

The organization of the storage hierarchy in Linux bears some resemblance to the network stack in Linux. Both are multi-layered and strictly define the role of each layer in the stack. Device drivers and physical interfaces are involved that dictate the overall performance. Similar to the behavior of the block layer, when a network packet was ready for transmission, it was placed in a single queue. This approach was used for several years until the network hardware evolved to support multiple queues. Hence, for devices with multiple queues, this approach became obsolete.

This problem was pretty similar to the one that was later faced by the block layer in the kernel. The network stack in the Linux kernel solved this problem a lot earlier than the storage stack. Hence, the kernel's storage stack took a cue from this, which led to the creation of a new framework for the Linux block layer, known as the **multi-queue block** I/O queuing mechanism, shortened to **blk-mq**.

The multi-queue framework solved the limitations in the block layer by isolating request queues for every CPU core. *Figure 5.2* illustrates how this approach fixes all three limitations in the single queue framework's design:

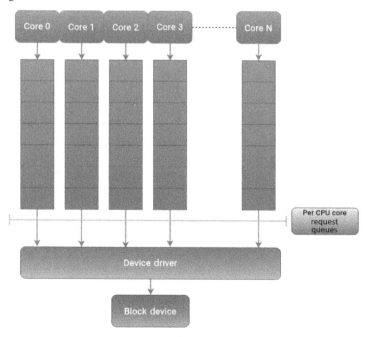

Figure 5.2 – The multi-queue framework

By using this approach, a CPU core can focus on executing its threads without worrying about the threads running on other cores. This approach resolves the limitations caused by the shared global lock and also minimizes the usage of interrupts and the need for cache coherency.

The `blk-mq` framework implements the following two-level queue design for handling I/O requests:

- **Software staging queues**: The software staging queues that are represented consist of one or more `bio` structures. A block device will have multiple software I/O submission queues, usually one per CPU core, and each queue will have a lock. A system with $M$ sockets and $N$ cores can have a minimum of $M$ and a maximum of $N$ queues. Each core submits I/O requests in its queue and doesn't interact with other cores. These queues eventually fan into a single queue for the device driver. The I/O schedulers can operate on the requests in the staging queue to reorder or merge them. However, this reordering doesn't matter as SSDs and NVMe drives don't care if an I/O request is random or sequential. This scheduling happens only between requests in the same queue, so no locking mechanism is required.

- **Hardware dispatch queues**: The number of hardware queues that can be supported depends on the number of hardware contexts that are supported by the hardware and its corresponding device driver. However, it should be noted that the maximum number of hardware queues will not exceed the number of cores in the system. The number of software staging queues can be less than, greater than, or equal to the number of hardware queues. The hardware dispatch queues represent the last stage of the block layer's code and act as a mediator before the requests get handed over to the device driver for their final execution. When an I/O request arrives at the block layer and there isn't an I/O scheduler associated with the block device, `blk-mq` will send the request directly to the hardware queue.

The multi-queue API makes use of tags to indicate which request has been completed. Every request is identified by a tag, which is an integer value ranging from zero to the size of the dispatch queue. The block layer generates a tag, which is subsequently utilized by the device driver, eliminating the need for a duplicate identifier. Once the driver has finished processing the request, the tag is returned to the block layer to signal the completion of the operation. The following section highlights some of the major data structures that play a vital role in the implementation of the multi-queue block layer.

## Looking at data structures

Here are some of the primary data structures that are essential to implement the multi-queue block layer:

- The first relevant data structure that's used by the multi-queue framework is the `blk_mq_register_dev` structure, which contains all the necessary information required when registering a new block device to the block layer. It contains various fields that provide details about the driver's capabilities and requirements.

- The `blk_mq_ops` data structure serves as a reference for the multi-queue block layer to access the device driver's specific routines. This structure serves as an interface for communication between the driver and the `blk-mq` layer, enabling the driver to integrate seamlessly into the multi-queue processing framework.

- The software staging queues are represented by the `blk_mq_ctx` structure. This structure is allocated on a per-CPU core basis.

- The corresponding structure for hardware dispatch queues is defined by the `blk_mq_hw_ctx` struct. This represents the hardware context with which a request queue is associated.

- The task of mapping software staging queues to hardware dispatch queues is performed by the `blk_mq_queue_map` structure.

- The requests are created and sent to the block device through the `blk_mq_submit_bio` function.

The following figure paints a picture of how these functions are interconnected:

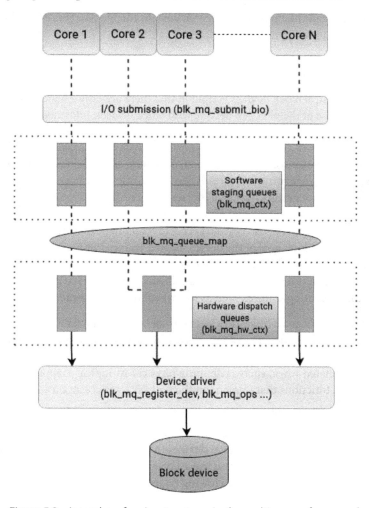

Figure 5.3 – Interplay of major structures in the multi-queue framework

To summarize, the multi-queue interface solves the limitations faced by the block layer when working with modern storage devices that have multiple queues. Historically, regardless of the capabilities of the underlying physical storage medium, the block layer maintained a single-request queue to handle I/O requests. On systems with multiple cores, this quickly turned into a major bottleneck. As the request queue was being shared between all CPU cores through a global lock, a considerable amount of time was spent by each CPU core waiting for the lock to be released by another core. To overcome this challenge, a new framework was developed to cater to the requirements of modern processors and storage devices. The multi-queue framework resolves the limitations of the block layer by segregating request queues for each CPU core. This framework leverages a dual queue design that is comprised of software staging queues and hardware dispatch queues.

With that, we have analyzed the multi-queue framework in the block layer. We will now shift our focus and explore the device mapper framework.

## Looking at the device mapper framework

By default, managing physical block devices is rigid in that there are only a handful of ways in which an application can make use of them. When dealing with block devices, informed decisions have to be made regarding disk partitioning and space management to ensure optimal usage of available resources. In the past, features such as thin provisioning, snapshots, volume management, and encryption were exclusive to enterprise storage arrays. However, over time, these features have become crucial components of any local storage infrastructure. When operating with physical drives, it is expected that the upper layers of the operating system will possess the necessary capabilities to implement and sustain these functionalities. The Linux kernel provides the device mapper framework for implementing these concepts. The device mapper is used by the kernel to map physical block devices to higher-level virtual block devices. The primary goal of the device mapper framework is to create a high-level layer of abstraction on top of physical devices. The device mapper provides a mechanism to modify bio structures in transit and map them to block devices. The use of the device mapper framework lays the foundation for implementing features such as logical volume management.

The device mapper provides a generic way to create virtual layers of block devices on top of physical devices and implement features such as striping, mirroring, snapshots, and multipathing. Like most things in Linux, the functionality of the device mapper framework is divided into kernel space and user space. The policy-related work, such as defining physical-to-logical mappings, is contained in the user space, while the functions that implement the policies to establish these mappings lie in the kernel space.

The device mapper's application interface is the `ioctl` system call. This system call adjusts the special file's underlying device parameters. The logical devices that employ the device mapper framework are managed via the `dmsetup` command and the `libdevmapper` library, which implement the respective user interface, as depicted in the following figure:

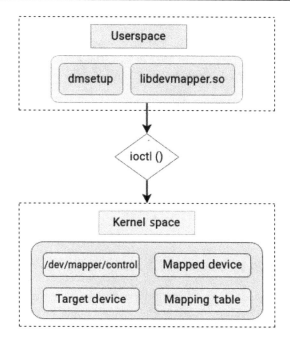

Figure 5.4 – Major components of the device mapper framework

If we run `strace` on the `dmsetup` command, we will see that it makes use of the `libdevmapper` library and the `ioctl` interface:

```
root@linuxbox:~# strace dmsetup ls
execve("/sbin/dmsetup", ["dmsetup", "ls"], 0x7fffbd282c58 /* 22 vars
*/) = 0
[................]
access("/etc/ld.so.nohwcap", F_OK)       = -1 ENOENT (No such file or
directory)
openat(AT_FDCWD, "/lib/x86_64-linux-gnu/libdevmapper.so.1.02.1", O_
RDONLY|O_CLOEXEC) = 3
[...............…]
stat("/dev/mapper/control", {st_mode=S_IFCHR|0600, st_rdev=makedev(10,
236), ...}) = 0
openat(AT_FDCWD, "/dev/mapper/control", O_RDWR) = 3
openat(AT_FDCWD, "/proc/devices", O_RDONLY) = 4
[...............…]
ioctl(3, DM_VERSION, {version=4.0.0, data_size=16384, flags=DM_EXISTS_
FLAG} => {version=4.41.0, data_size=16384, flags=DM_EXISTS_FLAG}) = 0
ioctl(3, DM_LIST_DEVICES, {version=4.0.0, data_size=16384, data_
start=312, flags=DM_EXISTS_FLAG} => {version=4.41.0, data_size=528,
data_start=312, flags=DM_EXISTS_FLAG, ...}) = 0
[................]
```

Applications that establish mapped devices, such as LVM, communicate with the device mapper framework via the `libdevmapper` library. The `libdevmapper` library utilizes `ioctl` commands to transmit data to the `/dev/mapper/control` device. The `/dev/mapper/control` device is a specialized device that functions as a control mechanism for the device mapper framework.

From *Figure 5.4*, we can see that the device mapper framework in kernel space implements a modular architecture for storage management. The device mapper framework's functionality consists of the following three major components:

- Mapped device
- Mapping table
- Target device

Let's briefly look at their respective roles.

### Looking at the mapped device

A block device, such as a whole disk or an individual partition, can be *mapped* to another device. The mapped device is a logical device provided by the device mapper driver and usually exists in the `/dev/mapper` directory. Logical volumes in LVM are examples of mapped devices. The mapped device is defined in `drivers/md/dm-core.h`. If we look at this definition, we will come across a familiar structure:

```
struct mapped_device {
[........]
struct gendisk *disk;
[..........]
```

The `gendisk` structure, as explained in *Chapter 4*, represents the notion of a physical hard disk in the kernel.

### Looking at the mapping table

A mapped device is defined by a mapping table. This mapping table represents a mapping from a mapped device to target devices. A mapped device is defined by a table that describes how each range of logical sectors of the device should be mapped, using a device table mapping that is supported by the device mapper framework. The mapping table defined in `drivers/md/dm-core.h` contains a pointer to the mapped device:

```
struct dm_table {
        struct mapped_device *md;
[...............]
```

This structure allows mappings to be created, modified, and deleted in the device mapper stack. Details about the mapping table can be viewed by running the `dmsetup` command.

## *Looking at the target device*

As explained earlier, the device mapper framework creates virtual block devices by defining mappings on physical block devices. Logical devices are created using "targets," which can be thought of as modularized plugins. Different mapping types, such as linear, mirror, snapshot, and others, can be created using these targets. Data is passed from the virtual block device to the physical block device through these mappings. The target device structure is defined in include/linux/device-mapper.h. The unit that's used for mapping is a sector:

```
struct dm_target {
        struct dm_table *table;
        sector_t begin;
        sector_t len;
[.............]
```

The device mapper can be a bit confusing to understand, so let's illustrate a simple use case of the building blocks that we explained previously. We're going to use the *linear* target, which lays the foundation of logical volume management. As discussed earlier, we're going to use the dmsetup command for this purpose as it implements the user-space functionality of the device mapper. We're going to create a linear mapping target called dm_disk. If you plan on running the following commands, make sure that you run them on a blank disk. Here, I've used two disks, sdc and sdd (you can use any disk for the exercise, so long it's empty!). Note that once you press *Enter* after the dmsetup create commands, it will prompt you for input. The sdc and sdd disks are referred to using their respective major and minor numbers. You can find out the major and minor numbers for your disk using lsblk. The major and minor numbers for sdc are 8 and 32, expressed as 8:32. Similarly, for sdd, this combination is expressed as 8:48. The rest of the input fields will be explained shortly. Once you've entered the required data, use *Ctrl + D* to exit. The following example will create a linear target of 5 GiB:

```
[root@linuxbox ~]# dmsetup create dm_disk
dm_disk: 0 2048000 linear 8:32 0
dm_disk: 2048000 8192000 linear 8:48 1024
[root@linuxbox ~]#
[root@linuxbox ~]# fdisk -l /dev/mapper/dm_disk
Disk /dev/mapper/dm_disk: 4.9 GiB, 5242880000 bytes, 10240000 sectors
Units: sectors of 1 * 512 = 512 bytes
Sector size (logical/physical): 512 bytes / 512 bytes
I/O size (minimum/optimal): 512 bytes / 512 bytes
[root@linuxbox ~]#
```

Here's what we've done:

1.  We have created a logical device called dm_disk by using specific portions or ranges from two physical disks, sdc and sdd.

2.  The first line of input that we've entered, dm_disk: 0 2048000 linear 8:32 0, means that the first 2048000 sectors (0-2047999) of dm_disk will use the sectors of /dev/sdc, starting from sector 0. Therefore, the first 2048000 (0-2047999) sectors of sdc will be used by dm_disk.

3.  The second line, dm_disk: 2048000 8192000 linear 8:48 1024, means that the next 8192000 sectors (after sector number 2047999) of dm_disk are being allocated from sdd. These 8192000 sectors from sdd will be allocated from sector number 1024 onward. If the disks do not contain any data, we can use any sector number here. If existing data is present, then the sectors should be allocated from an unused range.

4.  The total number of sectors in dm_disk will be 8192000 + 2048000 = 10240000.

5.  With a sector size of 512 bytes, the size of dm_disk will be (8192000 x 512) + (2048000 x 512) ≈ 5 GiB.

The 0-2047999 sector numbers of dm_disk are mapped from sdc, whereas the 2048000-10239999 sector numbers are mapped from sdd. The example we've discussed is a simple one, but it should be evident that we can map a logical device to any number of drives and implement different concepts.

The following figure summarizes what we explained earlier:

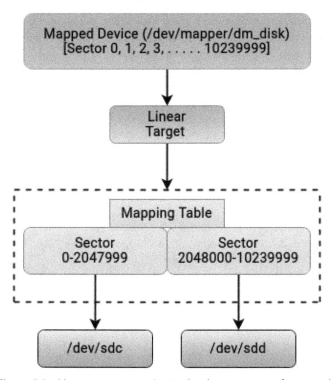

Figure 5.5 – Linear target mapping in the device mapper framework

The device mapper framework supports a wide variety of targets. Some of them are explained here:

- **Linear**: As we saw earlier, a linear mapping target can map a continuous range of blocks to another block device. This is the basic building block of logical volume management.

- **Raid**: The raid target is used to implement the concept of software raid. It is capable of supporting different raid types.

- **Crypt**: The crypt target is used to encrypt data on block devices.

- **Stripe**: The stripe target is used to create a striped device called (`raid 0`) across multiple underlying disks.

- **Multipath**: The multipath mapping target is utilized in storage environments where a host has multiple paths to a storage device. It allows a multipath device to be mapped.

- **Thin**: The thin target is used for thin provisioning – that is, creating devices larger than the size of the underlying physical device. The physical space is allocated only when written to.

As repeatedly mentioned earlier, the linear mapping target is most commonly implemented in LVM. Most Linux distributions use LVM by default for space allocation and partitioning. To the common user, LVM is probably one of the more well-known features of Linux. It should not be too difficult to see how the previously mentioned example can be applied to LVM or any other target for that matter.

As most of you should be aware, LVM is divided into three basic entities:

- **Physical volume**: The physical volume is at the lowest layer. The underlying physical disk or partition is a physical volume.

- **Volume group**: The volume group divides the space available in a physical volume into a sequence of chunks, called physical extents. A physical extent represents a contiguous range of blocks. It is the smallest unit of disk space that can be individually managed by LVM. By default, an extent size of 4 MB is used.

- **Logical volume**: From the space available in a volume group, logical volumes can be created. Logical volumes are typically divided into smaller chunks of data, each known as a logical extent. Since LVM utilizes linear target mapping, there is a direct correspondence between physical and logical extents. Consequently, a logical volume can be viewed as a mapping that's established by LVM that associates logical extents with physical ones. This can be visualized in the following figure:

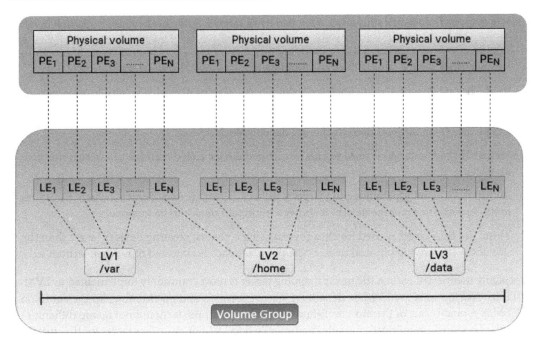

Figure 5.6 – LVM architecture

The logical volumes, as we all know, can be treated like any regular block device, and filesystems can be created on top of them. A single logical volume that spans multiple physical disks is similar to RAID-0. The type of mapping to be used between physical and logical extents is determined by the target. As LVM is based on a linear target, there is a one-to-one mapping relationship between physical and logical extents. Let's say we were to use the dm-raid target and configure RAID-1 to do mirroring between multiple block devices. In that case, multiple physical extents will map to a single logical extent.

Let's wrap up our discussion of the device mapper framework by mapping some key facts in our minds. The device mapper framework plays a vital role in the kernel and is responsible for implementing several key concepts in the storage hierarchy. The kernel uses the device mapper framework to map physical block devices to higher-level virtual block devices. The functionality of the device mapper framework is split into user space and kernel space. The user-space interface consists of the libdevmapper library and the dmsetup utility. The kernel part consists of three major components: the mapped device, mapping table, and target device. The device mapper framework provides the basis for several important technologies in Linux, such as LVM. LVM provides a thin layer of abstraction above physical disks and partitions. This abstraction layer allows storage administrators to easily resize filesystems based on their space requirements, providing them with a high level of flexibility. Before concluding this chapter, let's briefly touch on the caching mechanisms that are employed by the block layer.

# Looking at multi-tier caching mechanisms in the block layer

The performance of physical storage is usually orders of magnitude slower than that of processors and memory. The Linux kernel is well aware of this limitation. Hence, it uses the available memory as a cache and performs all operations in memory before writing all data to the underlying disks. This caching mechanism is the default behavior of the kernel and it plays a central role in improving the performance of block devices. This also positively contributes toward improving the system's overall performance.

Although solid state and NVMe drives are now commonplace in most storage infrastructures, the traditional spinning drives are still being used for cases where capacity is required and performance is not a major concern. When we talk about drive performance, random workloads are the *Achilles heel* of spinning mechanical drives. In comparison, the performance of flash drives does not suffer from such limitations, but they are far more expensive than mechanical drives. Ideally, it would be nice to get the advantages of both media types. Most storage environments are hybrid and try to make efficient use of both types of drives. One of the most common techniques is to place *hot* or frequently used data on the fastest physical medium and move *cold* data to slower mechanical drives. Most enterprise storage arrays offer built-in storage tiering features that implement this caching functionality.

The Linux kernel is also capable of implementing such a cache solution. The kernel offers several options to combine the capacity offered by spinning mechanical drives with the speed of access offered by SSDs. As we saw earlier, the device mapper framework offers a wide variety of targets that add functionalities on top of block devices. One such target is the dm-cache target. The dm-cache target can be used to improve the performance of mechanical drives by migrating some of its data to faster drives, such as SSDs. This approach is a bit contrary to the kernel's default caching mechanism, but it can be of significant use in some cases.

Most cache mechanisms offer the following operational modes:

- **Write-back**: This mode caches newly written data but does not write it immediately to the target device.

- **Write-through**: In this mode, new data is written to the target while still retaining it in the cache for subsequent reads.

- **Write-around**: This mode implements read-only caching. Data written to the device goes directly to the slower mechanical drive and is not written to the fast SSD.

- **Pass-through**: To enable pass-through mode, the cache needs to be clean. Reading is served from the origin device that bypasses the cache. Writing is forwarded to the origin device and *invalidates* the cache block.

The dm-cache target supports all the previously mentioned modes, except write-around. The required functionality is implemented through the following three devices:

- **Origin device**: This will always be the slow primary storage device

- **Cache device**: This is a high-performing drive, usually an SSD

- **Metadata device**: Although this is optional and this information can also be saved on the fast cache device, this device is used for keeping track of all the metadata information, such as which disk blocks are in the cache, which blocks are dirty, and so on

Another similar caching solution is dm-writecache, which is also a device mapper target. As its name suggests, the main focus of dm-writecache is strictly write-back caching. It only caches write operations and does not perform any read or write-through caching. The thought process for not caching reads is that read data should already be in the page cache. The write operations are cached on the faster storage device and then migrated to the slower disk in the background.

Another notable solution that has gained widespread popularity is bcache. The bcache solution supports all four caching modes defined previously. bcache uses a far more complex approach and lets all sequential operations go to the mechanical drives by default. Since SSDs excel at random operations, there generally won't be many benefits to caching large sequential operations on SSDs. Hence, bcache detects sequential operations and skips them. The writers for bcache compare it to the L2 **adaptive replacement cache** (**ARC**) in ZFS. The bcache project has also led to the development of the *Bcachefs* filesystem.

## Summary

This chapter was the second chapter in our exploration of the block layer in the kernel. The two main topics we discussed in detail were the multi-queue and device mapper frameworks. At the start of this chapter, we looked into the legacy single-request queue model in the block layer, its limitations, and its adverse impact on performance when working with modern storage drives and multi-core systems. From there, we introduced the multi-queue framework in the kernel. We described how the multi-queue framework addresses the limitations of the single-request model and improves the performance of modern storage drives, which are capable of supporting multiple hardware queues.

We also got a chance to look at the device mapper framework in the kernel. The device mapper framework is an essential part of the kernel and is responsible for implementing several technologies, such as multipathing, logical volumes, encryption, and raid. The most well known of these is logical volume management. We saw how the device mapper can implement these powerful features through mapping techniques.

In the next chapter, we'll conclude our discussion of the block layer after exploring the different I/O schedulers in the kernel.

# Understanding I/O Handling and Scheduling in the Block Layer

*"The key is not to prioritize what's on your schedule, but to schedule your priorities." – Stephen Covey*

*Chapter 4* and *Chapter 5* of this book focused on the role of the block layer in the kernel. We were able to see what constitutes a block device, the major data structures in the block layer, the multi-queue block I/O framework, and the device mapper. This chapter will focus on another important function of the block layer – scheduling.

Scheduling is an extremely critical component of any system, as the decisions taken by a scheduler can have a major say in dictating the overall system performance. The I/O scheduling in the block layer is no exception to this rule. The I/O scheduler holds significant importance in deciding the manner and timing of delivery for an I/O request to the lower layers. Given this, it becomes crucial to carefully analyze the I/O patterns of an application, as certain requests require prioritization over others.

This chapter will introduce us to the different I/O schedulers available in the block layer and their modus operandi. Each scheduler uses a different set of techniques to dispatch I/O requests to the lower layers. As we have mentioned repeatedly, when working with block devices, performance is a key concern. The block layer has gone through several enhancements so that maximum performance can be extracted from disk drives. This includes the development of schedulers to handle modern and high-performing storage devices.

We'll start by introducing the common techniques used by the different schedulers to handle I/O requests more efficiently. Although these techniques were developed for traditional spinning drives, they are still considered useful for modern flash drives. The primary goal of these techniques was to reduce disk-seeking operations for mechanical drives, as these have an adverse effect on their performance. Most schedulers make use of these methods by default, regardless of the underlying storage hardware.

The major topic of discussion in this chapter will be the different I/O scheduling flavors available in the kernel. The older disk schedulers were developed for devices accessed using the single-queue mechanism and have become outdated, as they cannot scale up to meet the performance of modern drives. In the last few years, four multi-queue I/O schedulers have been integrated into the kernel. These schedulers are able to map I/O requests to multiple queues.

In this chapter, we're going to discuss the following main topics:

- Understanding the I/O handling techniques in the block layer
- Explaining the I/O schedulers in Linux:
  - The MQ-deadline scheduler – guaranteeing a start service time
  - Budget fair queuing – providing proportional disk share
  - Kyber – prioritizing throughput
  - None – minimal scheduling overhead
- Discussing the scheduling conundrum

## Technical requirements

It would be helpful to have some knowledge about disk I/O basics to understand the concepts presented in this chapter. Having an idea about the different types of storage media, and concepts such as disk seek time and rotational latency, will help to comprehend the material presented in this chapter.

The commands and examples presented in this chapter are distribution-agnostic and can be run on any Linux operating system, such as Debian, Ubuntu, Red Hat, and Fedora. There are quite a few references to the kernel source code. If you want to download the kernel source, you can download it from `https://www.kernel.org`. The code segments referred to in this chapter and book are from the `5.19.9` kernel.

## Understanding the I/O handling techniques in block layer

While exploring the block layer in *Chapter 4* and *Chapter 5*, we often mentioned the performance sensitivity of block devices and how the block layer has to make informed and intelligent decisions to extract their maximum potential. So far, we haven't really discussed any of the techniques that help to enhance the performance of block devices.

Going back to the era of spinning drives, the performance of storage drives was a major bottleneck in the I/O stack. Mechanical drives offered decent performance when doing sequential I/O operations. However, for random workloads, their performance deteriorates quite drastically. This is understandable, as mechanical drives have to *seek* requested locations on disk by spinning and positioning the read-write head on specific locations. The greater the number of random seeks, the greater the performance penalty. Filesystems created on top of block devices try to implement some practices that attempt to optimize disk performance, but it is impossible to avoid random operations altogether.

Given the enormous seek times of mechanical drives, it is imperative that some sort of optimization is applied to reduce seeking, before I/O requests are handed over to the underlying storage. Simply handing them down to the underlying physical storage seems a primitive approach. This is where I/O schedulers come to the fore. The I/O schedulers in the block layer employ some common methods to ensure that the overhead caused by random access operations is minimized. These techniques address some of the performance issues of spinning drives, although they might not have a considerable effect when using flash drives, as they are not impacted by random operations.

Most schedulers employ a combination of the following techniques to optimize disk performance:

- Sorting
- Merging
- Coalescing
- Plugging

Let's discuss these in a bit more detail.

## Sorting

Let's say that four I/O requests, A, B, C, and D, are received for sectors 2, 3, 1, and 4, respectively, in that particular order, as illustrated in *Figure 6.1*. If the requests are delivered to the underlying spinning drive in this manner, they will be completed in that order:

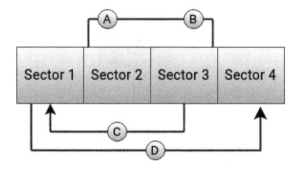

Figure 6.1 – Disk seeking

This means that after completing requests A and B in a sequential manner, for request C, the read-write head of the drive will have to go back to sector 1. After completing request C, it will have to perform another *seek* and move forward to sector D. It's not difficult to see the inefficiency that results from such an approach. If the requests are simply handed over in the received order, the disk performance will suffer.

For spinning drives, random access operations kill performance, as the disk has to perform multiple *seek* operations. If incoming requests are simply inserted at the end of a first-in-first-out queue, each request in the queue will involve separate processing, and the overhead caused by random seeking will increase. Therefore, most schedulers keep the request queues ordered and try to insert new incoming requests in a sorted manner. The request queue is sorted sector by sector. This ensures that requests operating on neighboring sectors can be performed sequentially.

## Merging

Merging acts as a compliment to the sorting mechanism and further tries to reduce random access. It can be performed in two ways, frontward and backward. Two requests can be merged if they are intended for contiguous sectors. If an I/O request enters the scheduler and it adjoins to an already enqueued request, then it qualifies as a front or back merge candidate. If the incoming request is merged with an existing request, it is called a back merge. The concept of back merging is shown in *Figure 6.2*:

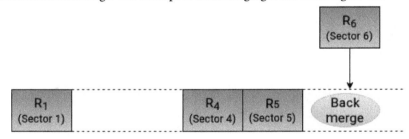

Figure 6.2 – A back merge

In the same vein, when a newly generated request is combined with an existing request, it is referred to as a front merge, as shown in *Figure 6.3*:

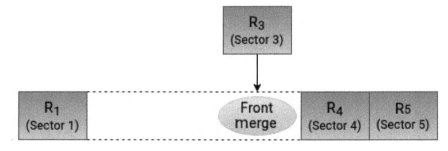

Figure 6.3 – A front merge

The idea is simple – avoid continuous trips to random locations. This is most effective for spinning mechanical drives. By default, most block layer schedulers attempt to merge an incoming request with an existing one.

## Coalescing

The coalescing operation includes both front and back merges. Coalescing happens when a new I/O request closes the gap between two existing requests, as shown in the following figure:

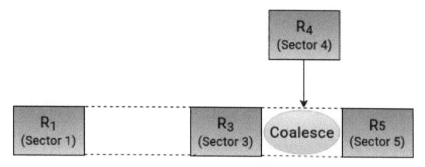

Figure 6.4 – Coalescing

Coalescing is employed to reduce the overhead associated with small and frequent I/O operations, particularly for spinning hard disk drives. By coalescing multiple requests, the disk can perform sequential reads and writes, resulting in faster I/O operations and reduced disk head movement.

## Plugging

The kernel uses the concept of *plugging* to stop requests from being processed in the queue. We're talking about improving performance here, so how does putting requests on hold help out? As we've learned, merging has a very positive effect on the drive performance. However, for smaller I/O requests to merge into a larger unified request, there must be existing requests for adjacent sectors in the queue. Therefore, in order to perform merging, the kernel first has to build up the request queue with a few requests so that there is a greater probability of merging. Plugging the queue helps to batch requests in anticipation of opportunities for merge and sort operations.

Plugging is a technique used to ensure that there are enough requests in the queue for potential merging operations. It involves waiting for additional requests to fill up the request queue and helps regulate the dispatch rate of requests to the device queue. The purpose of plugging is to control the dispatch rate of requests to the device queue. When there are no pending requests or a very small number of them in the block device queue, incoming requests are not dispatched to the device driver immediately. This results in the device being in a plugged state. The following figure demonstrates this concept:

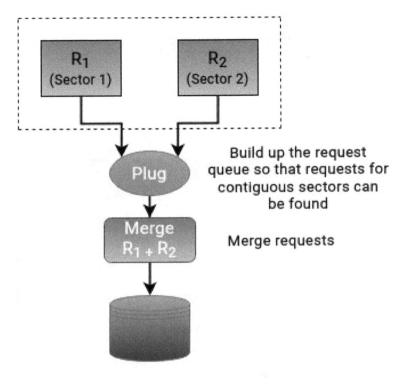

Figure 6.5 – Plugging

Plugging is executed at the process level rather than at the device level. The kernel initiates a plug sequence when a process carries out I/O operations. After the process has completed submitting its I/O requests to the queue, the requests are forwarded to the block layer and then dispatched to the device driver. A device is considered unplugged once the process has finished submitting I/O requests. If an application is blocked during a plug sequence, the scheduler proceeds to process the requests that are already in the queue.

Having discussed the most commonly used I/O scheduling techniques found in I/O schedulers, let us now delve into the reasoning and principles behind the decision-making process of the most widely used I/O schedulers in Linux.

## Explaining the Linux I/O schedulers

Disk schedulers are an interesting topic. They serve as a bridge between the block layer and low-level device drivers. The requests issued to a block device are altered by an I/O scheduler and handed over to the device drivers. It is the job of the scheduler to perform operations such as merging, sorting, and plugging on the I/O requests and divide the storage resources among the queued I/O requests. One of the notable advantages of the disk schedulers in Linux is their Plug and Play capability, allowing them to be switched in real time. Additionally, depending on the characteristics of the storage hardware being used, a distinct scheduler can be assigned to each block device in the system. The selection of a disk scheduler is not something that frequently comes under the radar, unless you're trying to extract the maximum from your system. The I/O scheduler is in charge of deciding the order in which I/O requests will be delivered to the device driver. The order is decided with a priority on the following tasks:

- Reducing disk seeking
- Ensuring fairness among I/O requests
- Maximizing disk throughput
- Reducing latency for time-sensitive tasks

It's a strenuous task to strike a balance among these goals. The different schedulers make use of multiple queues to achieve these goals. Operations such as merging and sorting are performed in request queues. The schedulers also perform additional processing, as per their internal algorithms, in these queues. Once the requests are ready, they are handed over to the dispatch queue managed by the device drivers.

The major performance optimizations in earlier block layer designs were directed toward hard disk drives. This is especially true for the disk scheduling algorithms. Most of the I/O handling techniques that we've discussed so far are most useful when the underlying storage media consists of rotating mechanical drives. As we'll see in *Chapter 7*, SSDs and NVMe drives are different beasts of nature and are not impacted by the limitations that impede a mechanical drive.

The scheduler controls the behavior of the underlying disks and thus plays a vital role in dictating the performance of an application. Just like the varying natures of physical storage, every application is also built differently. It is imperative to know the type of workload for the environment being tuned. There is no single scheduler that can be deemed fit enough to match the varying I/O characteristics of all applications. When choosing a scheduler, it is crucial to ask the following questions:

- What is the host system type – that is, is it a desktop, laptop, virtual machine, or a server?
- What sort of workload will be running? What type of application? Databases, multi-user desktop interface, games, or videos?
- Is the hosted application a processor or I/O-bound?
- What is the backend storage media type? HDDs, SSDs, or NVMe?
- Is the storage local to the host? Or is it provisioned from a large enterprise storage area network?

The I/O requests generated by a real-time application should be completed within a certain deadline. For instance, when streaming a video through a multimedia player, it has to be guaranteed that the frames will be read in time so that the video can be played without any glitches. On the other hand, interactive applications have to wait for the completion of a task before proceeding to the next one. For example, when writing in a document editor, the end user expects the editor to respond immediately when a key is pressed. Plus, the text has to appear in the same order in which it was typed.

For individual systems, the choice of a scheduler may not matter much and default settings might suffice. For servers running enterprise workloads, margins are much finer, and how a scheduler handles I/O requests may well decide the overall performance of the application. As we have repeatedly mentioned in this book, disk I/O is much slower than the processor and memory subsystems. Therefore, any decision regarding the choice of a disk scheduler should be accompanied by a lot of consideration and performance benchmarking.

Disk scheduling should not be confused with CPU scheduling. To process any request, both I/O and CPU time are required. In simpler terms, a process requests time from the CPU, after which it is able to run (if the time is allocated). The process can issue read or write requests to the disk. It is then the job of the disk scheduler to order and shepherd those requests to the underlying disks.

I/O schedulers in Linux are also referred to as elevators. The elevator algorithm, also called SCAN, compares the operation of legacy mechanical drives with elevators or lifts. When elevators go either up or down, they keep going in that same direction and stop to drop off people along the way. In disk scheduling, the read-write head of the drive starts from one end of the disk and moves toward the other end, while servicing requests along the way. To continue the analogy, mechanical drives need to read (pick up) and write (drop off) requests (people) at different disk locations (floors).

The different types of I/O schedulers available in the kernel are suited to particular use cases, some more than others. As we learned in *Chapter 5*, the single-queue framework does not scale up to meet the performance levels of modern storage devices. The advancements in drive technologies and multi-core systems led to the development of the multi-queue block I/O framework. Even with the implementation of this framework, the kernel still lacked an important ingredient when dealing with modern drives – an I/O scheduler designed to work with multi-queue devices. Schedulers that were designed for the single-queue framework and intended to be used with single-queue devices do not function optimally with modern drives.

*Figure 6.6* highlights the various types of I/O schedulers available for both single and multi-queue frameworks:

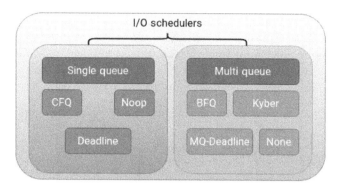

Figure 6.6 – Different I/O Scheduling options

Single-queue I/O schedulers are deprecated and have not been a part of the kernel since version 5.0. Although you can disable these and revert back to the single-queue schedulers, the latest kernel releases default to the multi-queue schedulers, and as such we will keep our focus on the multi-queue schedulers that are a part of the kernel. There are four major players in this category. These schedulers map I/O requests to multiple queues, which are handled by kernel threads distributed across the multiple CPU cores:

- MQ-deadline
- **Budget Fair Queuing (BFQ)**
- Kyber
- None

Let's take a look at the operational logic of these schedulers.

## The MQ-deadline scheduler – guaranteeing a start service time

The deadline scheduler, as the name suggests, imposes a deadline to service I/O requests. Due to its latency-oriented design, it is often used for latency-sensitive workloads. Because of its high performance, it has also been adopted for multi-queue devices. Its implementation for multi-queue devices is known as *mq-deadline*.

The primary objective of the deadline scheduler is to ensure that a request has a designated start service time. This is accomplished by enforcing a deadline on all I/O operations, which helps prevent requests from being neglected. The deadline scheduler makes use of the following queues:

- **Sorted**: The read and write operations in this queue are sorted by the sector numbers they are to access.
- **Deadline**: The deadline queue is a standard **First-In-First-Out** (**FIFO**) queue that contains requests sorted by their deadlines. To prevent starvation of requests, the deadline scheduler utilizes separate instances of the deadline queue for read and write requests, assigning an expiration time to each I/O request.

The deadline scheduler places each I/O request in both the sorted and deadline queues. Before deciding which request to serve, the deadline scheduler selects a queue from which to choose a read or write request. If there are requests in both the read and write queues, read queues are preferred. This is because write requests can starve read operations. This makes the deadline scheduler extremely effective for read-heavy workloads.

The operational logic of the deadline scheduler is depicted in the following figure:

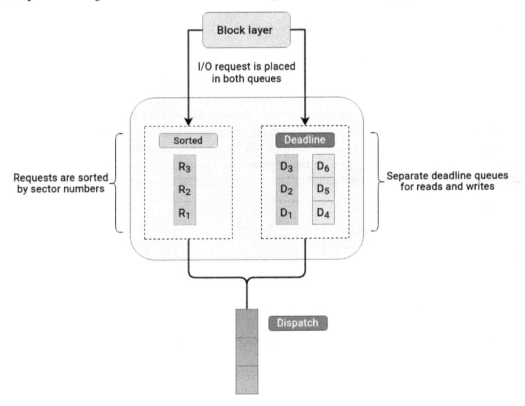

Figure 6.7 – The MQ-deadline I/O scheduler

The I/O requests to be served are decided as follows:

1.  Let's say that the scheduler has decided to serve read requests. It will check the first request in the deadline queue. If the timer associated with that request has expired, it will be handed over to the dispatch queue and inserted at its tail end. The scheduler then turns its focus to the sorted queue and selects a batch of requests (16 requests by default) following the chosen request. This is done to increase the sequential operations. Think of how an elevator drops off people on different floors along the way to its final destination. The number of requests in a batch is a tunable parameter and can be changed.

2.  It can also happen that there are no requests with expired deadlines in the deadline queue. In that case, the scheduler will examine the last request that was serviced from the sorted queue and choose the subsequent request in the sequence. The scheduler will then select a batch of 16 requests that follow the chosen request.

3.  After processing each batch of requests, the deadline scheduler checks to see whether requests in the write deadline queue have been starved for too long, and then decides whether to start a new batch of read or write operations.

The following figure explains this process. If a read request for sector number 19 on the disk is received, it is assigned a deadline and inserted at the tail end of the deadline queue for read operations. Based on the sector number, this request is also placed in the sorted sector queue, just behind the request for sector 11. The operational flow of the deadline scheduler, regarding how requests are processed, is demonstrated in *Figure 6.8*:

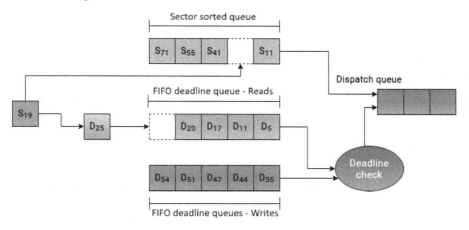

Figure 6.8 – Request handling in the MQ-deadline I/O scheduler

The following snippet code of mq-deadline in `block/mq-deadline` dictates some of the behavior illustrated in the preceding figure. The expiry deadline for read requests (HZ/2) is 500 milliseconds, whereas for writes, it is 5 seconds (5*HZ). This ensures that read requests have higher precedence. The term HZ represents the clock ticks generated per second. The definition of `writes_starved` indicates that reads can starve writes. The writes are only serviced once against two rounds of reads. `fifo_batch` sets the number of requests that can be batched together:

```
[..........]
static const int read_expire = HZ / 2;
static const int write_expire = 5 * HZ;
static const int writes_starved = 2;
static const int fifo_batch = 16;
[..........]
```

To summarize, the deadline scheduler strives to reduce I/O latency by implementing start service times for every incoming request. Each new request is assigned a deadline timer. When the expiry time for a request is reached, the scheduler will forcefully service that request to prevent request starvation.

## Budget fair queuing – providing proportional disk share

The **Budget Fair Queuing** (**BFQ**) scheduler is a relative newcomer in the world of disk schedulers, but it has gained considerable popularity. It is modeled after the **Completely Fair Queuing** (**CFQ**) scheduler. It provides fairly good response times and is considered particularly suitable for slower devices. With its rich and comprehensive scheduling techniques, the BFQ is often thought to be one of the most complete disk schedulers, although its sophisticated design also makes it the most complex scheduler among the lot.

BFQ is a proportional share disk scheduler. The primary goal of BFQ is to be fair to all I/O requests. To achieve this fairness, it makes use of some intricate techniques. Internally, BFQ uses the `Worst-case Fair Weighted Fair Queuing+ (B-WF2Q+)` algorithm to aid in scheduling decisions.

The BFQ scheduler guarantees a proportional share of the disk resources to every process in the system. It collects the I/O requests in the following two queues:

- **Per-process queues**: The BFQ scheduler allocates a queue for every process. Each per-process queue contains synchronous I/O requests.

- **Per-device queue**: All the asynchronous I/O requests are collected in a per-device queue. This queue is shared among processes.

Whenever a new queue is created, it is assigned a variable budget. Unlike most schedulers, which allocate time slices, this budget is implemented as the number of sectors that each process is allowed to transfer when it is next scheduled to access the disk resources. The value of this budget is what ultimately determines the share of disk throughput for each process. As such, its calculation is complex and based on a multitude of factors. The major factors in this calculation are the I/O weight and the recent I/O activity of the process. Based on these observations, the scheduler assigns a budget that is proportional to a process's I/O activity. The I/O weight of a process has a default value, but it can be changed. The assignment of the budget is such that a single process is not able to hog all the bandwidth of available storage resources. *Figure 6.9* shows the different queues used by the BFQ scheduler:

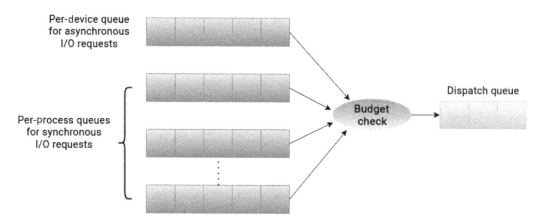

Figure 6.9 – The BFQ I/O scheduler

When it comes to servicing I/O requests, some of the factors affecting scheduling decisions are described as follows:

- The BFQ scheduler selects the queue to be served through the C-LOOK algorithm. It picks up the first request from the selected queue and hands it to the driver. The budget of the queue gets decremented by the size of the request This is explained in a bit more detail at the end of this discussion. BFQ exclusively serves one queue at a time.

- The BFQ scheduler prioritizes scheduling processes that have smaller I/O budgets. Normally, these are the processes that have a small but random set of I/O requests. In contrast, I/O intensive processes with a large number of sequential I/O requests are assigned a larger budget. When selecting a process queue for servicing, the BFQ scheduler chooses the queue with the lowest I/O budget, granting exclusive access to the disk resources. This approach achieves two objectives. First, processes with smaller budgets receive prompt service and do not have to wait excessively. Second, I/O-bound processes with larger budgets receive a proportionately greater share of disk resources, promoting sequential I/O operations and thereby enhancing disk performance. The BFQ scheduler makes use of a slightly unorthodox approach to increase disk throughput – performing disk idling by checking for synchronous I/O requests. When an application generates synchronous I/O requests, it enters a blocking state and waits for the operation to complete. Mostly, these are read requests, as write operations are asynchronous and can be directly performed in cache. If the last request in the process queue is synchronous, the process goes into a waiting state. This request is not dispatched immediately to the disk, as the BFQ scheduler allows the process to generate another request. During this time frame, the drive remains idle. More often than not, the process generates another request, as it waits for the current synchronous request to complete before issuing new requests. The new request is normally adjacent to the last request, which improves the chance of sequential operations. At times, this approach can backfire and might not always have a positive impact on performance.

- If two processes work on neighboring areas on a disk, it makes sense to combine their requests so that sequential operations can be increased. In this case, BFQ merges the queues of both processes to enable the consolidation of requests. The incoming requests are compared with the next request of the in-service process, and if the two requests are close, the request queues for both processes are merged.

In the event that an application executing read requests depletes its queue while still having a surplus budget, the disk will be idled briefly to give that process a chance to issue another I/O request.

The scheduler continues to serve the queue until one of the following events occurs:

- The queue budget is exhausted
- All queue requests have been completed
- The idling timer expires while waiting for a new request from the process
- Too much time has been spent while servicing the queue

Upon examining the BFQ code found in `block/bfq-iosched.c`, you will discover a notable concept known as the **charge factor** for asynchronous requests:

```
static const int bfq_async_charge_factor = 3;
```

It was mentioned earlier that when a request to be serviced is selected from a queue, the budget of the queue is decremented by the size of the request – that is, the number of sectors in the request. This is true for synchronous requests, but for asynchronous requests, this cost is much higher. This is also one of the ways reads are prioritized over writes. For asynchronous requests, the queue is charged with the number of sectors in the request, multiplied by the value of `bfq_async_charge_factor`, which is three. According to the kernel documentation, the current value for the charge factor parameter was determined by following a tuning process that involved various hardware and software configurations.

In summary, the BFQ scheduler employs equitable queuing approaches by apportioning a proportion of the I/O throughput to each process. It makes use of per-process queues for synchronous requests and a per-device queue for asynchronous requests. It assigns a budget to each process. This budget is calculated based on the I/O priority and the number of sectors transferred by the process when it was scheduled the last time. Although the BFQ scheduler is complex and incurs a slightly larger overhead compared to other schedulers, it is widely used, as it improves system response times and minimizes latency for time-sensitive applications.

## Kyber – prioritizing throughput

The Kyber scheduler is also a relatively newer entry in the disk scheduling world. Although the BFQ scheduler is older than the Kyber scheduler, both officially became a part of the kernel version 4.12. The Kyber scheduler is specifically designed for modern high-performing storage devices.

Historically, the ultimate goal of disk schedulers has been to reduce seek times for mechanical drives so that the overhead caused by random access operations can be decreased. Consequently, the different disk schedulers have used complex and sophisticated techniques to achieve this common goal. Each scheduler prioritizes certain aspects of performance in varying ways, which introduces an additional overhead while processing I/O requests. As modern drives, such as SSDs and NVMe, are not hampered by random access operations, some of the complicated techniques used by certain schedulers might not apply to these devices. For instance, the BFQ scheduler has a slightly high overhead for each request, so it is not considered ideal for systems to have high throughput drives. This is where the Kyber scheduler comes in handy.

The Kyber scheduler doesn't have complex internal scheduling algorithms. It is intended to be used in environments that comprise high-performing storage devices. It uses a very straightforward approach and implements some basic policies to marshal I/O requests. The Kyber scheduler splits the underlying device into multiple domains. The idea is to maintain a queue for the different types of I/O requests. Upon inspecting the code found in `block/kyber-iosched.c`, we can observe the presence of the following request types:

```
[.......]
static const char *Kyber_domain_names[] = {
        [KYBER_READ] = "READ",
        [KYBER_WRITE] = "WRITE",
        [KYBER_DISCARD] = "DISCARD",
        [KYBER_OTHER] = "OTHER",
};
[.......]
```

The Kyber scheduler categorizes the requests as follows – reads, writes, discard, and other requests. The Kyber scheduler maintains queues for these types of requests. The discard request is used for devices such as SSDs. The filesystem on top of the device can issue this request to discard blocks not in use by the filesystem. For the type of request mentioned previously, the scheduler implements a limit on the corresponding number of operations in the device queue:

```
[......]
static const unsigned int Kyber_depth[] = {
        [KYBER_READ] = 256,
        [KYBER_WRITE] = 128,
        [KYBER_DISCARD] = 64,
        [KYBER_OTHER] = 16,
};
[......]
```

The crux of Kyber's scheduling approach is to limit the size of dispatch queues. This directly correlates with the time spent waiting for I/O requests in the request queue. The scheduler only sends a limited number of operations to the dispatch queue, which ensures that the dispatch queue is not too crowded. This results in the swift processing of the requests in the dispatch queue. Consequently, the I/O operations in the request queues don't have to wait too long to be serviced. This approach results in reduced latency. The following figure illustrates the logic of the Kyber scheduler:

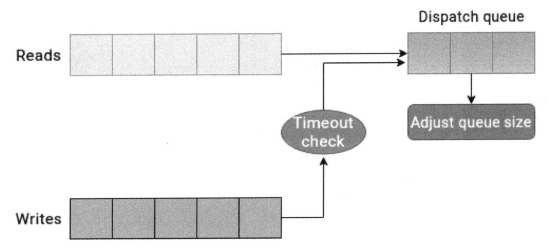

Figure 6.10 – The Kyber I/O scheduler

To determine the number of requests to be allowed in the dispatch queue, the Kyber scheduler uses a simple but effective approach. It calculates the completion time of each request, and based on this feedback, it adjusts the number of requests in the dispatch queue. Further, the target latencies for reads and synchronous writes are tunable parameters and can be changed. Based on their values, the scheduler will throttle requests in order to meet these target latencies.

The Kyber scheduler prioritizes requests in the read queue over those in the write queue, unless a write request has been outstanding for too long, meaning the target latency has been breached.

The Kyber scheduler is a performance powerhouse when it comes to modern storage devices. It is tailored for high-speed storage devices, such as SSDs and NVMe, and prioritizes low-latency I/O operations. This scheduler dynamically adjusts itself by scrutinizing I/O requests and enables the establishment of target latencies for both synchronous writes and reads. Consequently, it regulates I/O requests to meet the specified objectives.

# None – minimal scheduling overhead

The scheduling of I/O requests is a multifaceted problem. The scheduler has to take care of several aspects, such as reordering requests in the queue, allocating a portion of disk shares to each process, controlling the execution duration of every request, and making sure that individual requests do not monopolize the available storage resources. Each scheduler assumes that the host itself cannot optimize requests. Therefore, it jumps in and applies complex techniques to try and make the most of the available storage resources. The more sophisticated the scheduling technique, the greater the processing overhead. While optimizing requests, the schedulers generally make some assumptions about the underlying device. This works well unless the lower layers in the stack have better visibility of the available storage resources and can handle making scheduling decisions themselves, such as the following:

- In high-end storage settings, such as storage area networks, storage arrays frequently include their own scheduling logic, since they possess deeper insight into the nuances of the underlying devices. As a result, the scheduling of I/O requests typically transpires at the lower layer. When using raid controllers, the host system doesn't have complete knowledge about the underlying disks. Even if the scheduler applies some optimizations to I/O requests, it might not make much of a difference, as the host system lacks the visibility to accurately re-order the requests to lower seek time. In such cases, it makes sense to simply dispatch the requests to the raid controller.

- Most scheduler optimizations are directed toward slow mechanical drives. If the environment consists of SSDs and NVMe drives, the processing overhead associated with these scheduling optimizations may seem excessive.

In such cases, a unique but effective solution is to use the *none* scheduler. The none scheduler is the multi-queue *no-op I/O scheduler*. For single-queue devices, the same functionality was achieved through the *no-op* scheduler.

The none scheduler is the most straightforward of all schedulers, as it performs no scheduling optimizations. Every incoming I/O request is appended to a FIFO queue and delegated to the block device for handling. This strategy proves beneficial when it has been established that the host must not endeavor to rearrange requests according to their included sector numbers. The none scheduler has a single request queue that includes both read and write I/O requests. Due to its rudimentary approach, Although the none I/O scheduler imposes minimal overhead, it does not ensure any particular quality of service. The none scheduler also does not perform any reordering of requests. It only does request merging to reduce seek time and improve throughput. Unlike all the other schedulers, the none scheduler has no tunables or settings for optimization. The request merging operation is the entire extent of its complexity. Because of this, the none scheduler uses a minimal amount of CPU instructions per I/O request. The operation of the none scheduler is based on the assumption that devices at the lower layer, such as raid controllers or storage controllers, will optimize I/O performance.

The simple operational logic of the none scheduler is shown in *Figure 6.11*:

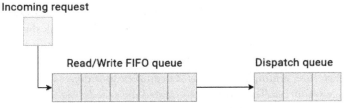

Figure 6.11 – The none I/O scheduler

Although every environment has a lot of variables, based on the mode of operation, the none scheduler seems to be the preferred scheduler for enterprise storage area networks, as it does not make any assumptions about the underlying physical devices, and it does not implement any scheduling decisions that can compete or clash with the logic of the lower level I/O controllers.

Given the profusion of options to choose from, it can be challenging to determine which scheduler is most suitable for your needs. In the subsequent section, we will outline common usage scenarios for the schedulers we have covered in this chapter.

## Discussing the scheduling conundrum

We've discussed and explained how the different I/O scheduling flavors go about their business, but the selection of a scheduler should always be accompanied by benchmark results gathered through real application workloads. As mentioned earlier, most of the time, default settings might be good enough. It's only when you try to achieve peak efficiency, you try and tinker with the default settings.

The pluggable nature of these schedulers means that we can change the I/O scheduler for a block device on the fly. There are two ways to do this. The currently active scheduler for a particular disk device can be checked through sysfs. In the following example, the active scheduler is set to mq-deadline:

```
[root@linuxbox ~]# cat /sys/block/sda/queue/scheduler
[mq-deadline] none bfq kyber
[root@linuxbox ~]#
```

To change the active scheduler, write the name of the desired scheduler to the scheduler file. For instance, to set the BFQ scheduler for sda, use the following command:

```
echo bfq > /sys/block/sda/queue/scheduler
```

The preceding method will only set the scheduler temporarily and revert to default settings after a reboot. To make this change permanent, edit the /etc/default/grub file and add the elevator=bfq parameter to the GRUB_CMDLINE_LINUX_DEFAULT line. Then, regenerate the GRUB configuration and reboot the system.

Merely changing the scheduler will not result in two-fold performance gains. Usually, the improvement figure will be somewhere between 10–20%.

Although each environment is different and scheduler performance may vary depending upon several variables, as a baseline, the following are some of the use cases of the schedulers that we've discussed in this chapter:

| Use case | Recommended I/O scheduler |
| --- | --- |
| A desktop GUI, interactive applications, and soft real-time applications, such as audio and video players | BFQ, as it guarantees good system responsiveness and low latency for time-sensitive applications |
| Traditional mechanical drives | BFQ or MQ-deadline – both are considered suitable for slower drives. Kyber/none are biased in favor of faster disks. |
| High-performing SSDs and NVMe drives as local storage | Preferably none, but Kyber might also be a good alternative in some cases |
| Enterprise storage arrays | None, as most storage arrays have built-in logic to schedule I/Os more efficiently |
| Virtualized environments | MQ-deadline is a good option. If the hypervisor layer does its own I/O scheduling, then using the none scheduler might be beneficial. |

Table 6.1 – Typical use cases for I/O schedulers

Please note that these are not strict use cases, as often, several conditions might be overlapping. The type of application, workload, host system, and storage media are just some of the factors that must be kept in mind before deciding on a scheduler. Typically, the deadline scheduler is regarded as a versatile choice, due to its modest CPU overhead. BFQ performs well in desktop environments, whereas none and Kyber are better suited for high-end storage devices.

## Summary

This chapter provided an overview of I/O scheduling, which is a critical function of the block layer. When a read or write request passes through all the layers of the virtual filesystem, it eventually arrives at the block layer. The chapter explored the various types of I/O schedulers and their characteristics, including their advantages and disadvantages. The block layer includes multiple I/O schedulers that are suitable for particular use cases. The choice of an I/O scheduler plays a vital role in determining how I/O requests will be handled at the lower layer. To make more performance-oriented decisions, most schedulers employ some common techniques that aid in improving overall disk performance. The techniques that we discussed in this chapter are merging, coalescing, sorting, and plugging.

We also explained the different scheduling options available in the kernel. The kernel has a separate set of I/O schedulers for single- and multi-queue devices. The single-queue schedulers have been deprecated since kernel version 5.0. The multi-queue scheduling options include the multi-queue Deadline scheduler, BFQ, Kyber, and the none scheduler. Each of these schedulers is suited to specific use cases, and there is no single recommendation, which can be applied to all situations. The MQ-deadline scheduler has good all-around performance. The BFQ scheduler is more oriented toward interactive applications, while Kyber and None are geared toward high-end storage devices. To choose a scheduler, it is imperative to know details about the environment, which includes details such as the type of workload, application, host system, and backend physical media.

This chapter concludes part two of the book, in which we delved into the block layer. In the next chapter, we'll see the different types of storage media available today and explain the differences between them.

# Part 3: Descending into the Physical Layer

This part will introduce you to the architecture and major components of the SCSI subsystem in the Linux kernel. You will also be introduced to the different types of physical storage media available today and the differences in their implementation.

This part contains the following chapters:

- *Chapter 7, The SCSI Subsystem*
- *Chapter 8, Illustrating the Layout of Physical Media*

# 7

# The SCSI Subsystem

Throughout this book, we've gradually traversed from the higher layers in the storage stack to the lower layer. We started from VFS, explored the major VFS structures and filesystems, and explored the structures and scheduling techniques in the block layer. The VFS and block layer represent a major portion of the software side of things in the I/O hierarchy. As we gradually move down the ladder and enter the physical layer, things are slightly more generic, as the lower-level standards used to address physical drives are the same for most systems.

The *third part* of this book contains two chapters that are dedicated to building an understanding of the physical side of things. In this chapter, we'll mainly focus on one particular subsystem that has existed for a while and is the most common standard and protocol for addressing physical devices, **Small Computer System Interface (SCSI)**.

The development of the SCSI protocol aimed to facilitate seamless data transfer between computers and peripheral devices, including disk drives, CD-ROMs, printers, scanners, and various other resources, ensuring efficient and reliable communication. Since we're focusing on storage here, we're only going to discuss its role in terms of disk drives. Any read or write request that is handed to SCSI from the upper layers is transformed into an equivalent SCSI command. It's important to understand that SCSI does not handle the arrangement of blocks for transportation or their physical placement on disk. That comes under the ownership of the upper layers in the I/O hierarchy.

Before we dive into SCSI, it is important to get a basic understanding of the device model in Linux. The **Linux Device Model**, based on the `kobject` structure in the kernel, provides a range of constructs that enable smooth communication and interaction between hardware devices and their corresponding device drivers. After getting some basic knowledge of the device model, we'll try to explain the major components of the SCSI subsystem.

In this chapter, we will discuss the following main topics:

- The device driver model
- The SCSI subsystem

# Technical requirements

As SCSI is a protocol used to communicate with peripheral devices, such as hard drives, some basic knowledge about the functioning of these devices will aid in understanding the SCSI subsystem.

The commands and examples presented in this chapter are distribution-agnostic and can be run on any Linux operating system, such as Debian, Ubuntu, Red Hat, or Fedora. There are quite a few references to the kernel source code. If you want to download the kernel source, you can download it from `https://www.kernel.org`.

# The device driver model

There are different subsystems in the kernel, such as the **system call interface, VFS, process and memory management**, and the **network stack**. Throughout this book, we've strictly kept our focus on the structures and entities that are a part of the I/O hierarchy in Linux. However, in reality, the process of reading and writing data to a storage device has to pass through most of these subsystems. As we saw, abstraction layers are the alpha and omega of the I/O stack, but this abstracted approach is not just limited to storage devices. For the kernel, the disk is just one of several pieces of hardware that it must manage. If there was a unique subsystem for managing the different types of devices, it would result in a bloated piece of code. Of course, different types of devices tend to be treated differently, as they might have contrasting roles, but for the end user, there should be a general abstract view of the system structure.

To achieve this unification, the Linux Device Model extracts the common attributes of device operations, abstracts them, and implements these common attributes in the kernel, providing a unified interface for newly added devices. This makes the driver development process much easier and smoother, as the developers only need to familiarize themselves with the interface.

The device model's main objective is to maintain internal data structures that accurately represent a system's state and configuration. This encompasses vital information such as the presence of devices, their associated buses and drivers, as well as the overall hierarchy and structure of buses, devices, and drivers within the system. To keep track of this information, the device model makes use of the following entities to map them to their physical counterparts:

- **Bus**: There are several components in a system, such as CPU, memory, and input and output devices. Communication between these devices is dependent on a channel, which is the bus. The bus is a channel to transmit data. Think of it as a linear channel for the conveyance of traffic, similar to a road. In order to facilitate the abstraction of the device model, all devices should be connected to a bus. The bus in the device model is an abstraction based on the physical bus.

- **Device**: This represents a physical device that is attached to a bus. In the device model, the *device* abstracts all hardware devices in the system and describes their attributes, the bus it is connected to, and other information.

- **Device driver**: The driver is a software entity that is associated with a device. The device model uses the driver to abstract the driver of the hardware device, which includes device initialization and power management-related interface implementations.

- **Class**: The concept of class is a bit interesting. A class represents a collection of devices with similar functions or attributes. For example, there are SCSI and **Advanced Technology Attachment** (**ATA**) drivers, which fall under the same disk class. Classes serve as a means of categorizing devices based on their functionality rather than their connectivity or operational mechanisms. This is a bit similar to the concept of classes in object-oriented programming.

The device model provides a generic mechanism to represent and operate on every device in the system. As we explained in *Chapter 1* of this book, the kernel provides a window to export information about various kernel subsystems through VFS. The representation of the device model in user space can be viewed through the `Sysfs` VFS. The `Sysfs` filesystem is mounted on the `/sys` directory, as shown in the following screenshot:

```
[root@linuxbox ~]# ls -l /sys/
total 0
drwxr-xr-x   2 root root 0 Feb  4 22:42 block
drwxr-xr-x  34 root root 0 Feb  4 22:50 bus
drwxr-xr-x  57 root root 0 Feb  4 22:41 class
drwxr-xr-x   4 root root 0 Feb  5 03:45 dev
drwxr-xr-x  16 root root 0 Jan  1 19:01 devices
drwxr-xr-x   7 root root 0 Jan  1 19:01 firmware
drwxr-xr-x   8 root root 0 Jan  1 19:01 fs
drwxr-xr-x   2 root root 0 Feb  5 01:26 hypervisor
drwxr-xr-x  16 root root 0 Jan  1 19:01 kernel
drwxr-xr-x 166 root root 0 Feb  5 03:45 module
drwxr-xr-x   2 root root 0 Feb  5 03:45 power
[root@linuxbox ~]#
```

Figure 7.1 – The contents of Sysfs

The directories contain the following information:

- `block`: This encompasses all available block devices within the system, including disks and partitions

- `bus`: This represents various types of buses to which physical devices are connected, such as PCI, IDE, and USB

- `class`: This denotes the available driver classes in the system, such as network, sound, and USB

- `devices`: This signifies the hierarchical structure of connected devices within the system

- `firmware`: This contains information retrieved from system firmware, particularly ACPI

- `fs`: This provides details regarding the mounted file systems

- `kernel`: This offers kernel status information, including logged-in users and hotplug events
- `module`: This presents the current list of loaded modules
- `power`: This encompasses information pertaining to the power management subsystem.

There is a correlation between the kernel data structures within the described model and the sub-directories in the `Sysfs` VFS. There are a number of structures in the device model that allow communication between a device driver and the corresponding hardware device. We're not going to explore these structures, but just so you know, the basic structure of the Linux Device Model is `kobject`. Think of `kobject` as the glue that holds the device model and the `Sysfs` interface together. The structures in the higher levels of the model, as depicted in *Figure 7.2*, are as follows:

- `struct bus_type`
- `struct device`
- `struct device_driver`

Here is *Figure 7.2*:

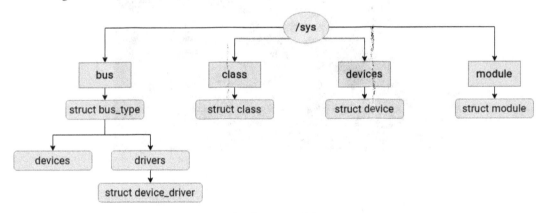

Figure 7.2 – The device model components

To summarize, the Linux device model classifies hardware devices and implements abstraction through a set of standard data structures and interfaces. This model can be seen through user space by viewing the contents of the `Sysfs` filesystem. The entities present in `Sysfs` have a close association with the actual physical implementations. Let's move on and explore the architecture of the SCSI subsystem in more detail.

# Explaining the SCSI subsystem

People can mean a couple of things when referring to the **SCSI** (pronounced SKUZ-ee):

- A hardware bus to connect peripherals to a computer

- A command set to communicate with devices over different types of buses

For a long time, SCSI was the primary technology for I/O buses in computers. SCSI defines both an interface and a data protocol for connecting different types of devices to a computer. As a medium, SCSI defines a bus for data transmission. As a protocol, it defines how devices communicate with each other via the SCSI bus.

Initially, the connectivity of peripheral devices was achieved through a parallel SCSI bus. Over the years, the SCSI parallel bus has fallen out of favor and has been replaced with serial interfaces. The most common of these interfaces include **Serial Attached SCSI (SAS)** and **SCSI over Fibre Channel**. The serial interfaces provide far better data transfer rates and reliability. There is also an implementation of the SCSI protocol over TCP/IP, known as **Internet SCSI (iSCSI)**.

We will keep our focus here on the Linux side of things and discuss the organization of the SCSI subsystem in the I/O hierarchy. The SCSI standard defines command sets for a wide variety of devices, not just for hard drives. The SCSI commands can be sent over just about any kind of transporting mechanism. This makes SCSI the ipso facto standard for storage devices accessed via the SATA, SAS, or Fibre Channel protocol.

## SCSI architecture

The SCSI subsystem uses a three-level architecture. The layer at the top represents the kernel's highest interface for end user applications. The layer at the center provides some common services to the upper and lower layers of the SCSI stack. At the very bottom is the lower layer. The lower layer contains the actual drivers that interact with the underlying physical devices. Every operation involving the SCSI subsystem uses one driver at each of three layers. *Figure 7.3* highlights the multilayered design of the SCSI subsystem:

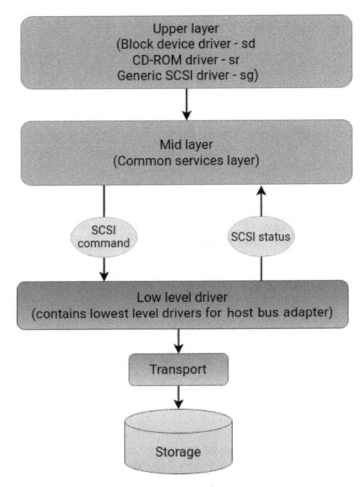

Figure 7.3 – SCSI architecture

The three layers are described in a bit more detail in the following subsections.

## *Upper layer*

The **upper layer** contains specific device-type drivers that are closest to the user space applications. These upper-layer drivers provide the interface between user space and kernel space. The most commonly used upper-layer drivers include the following:

- sd: The disk driver

- sr: The CD-ROM driver

- sg: The generic SCSI driver

After looking at these driver names, it should come as no surprise that the device names for the corresponding device types are abbreviated with the prefix of the driver, such as `sda`. The upper layer accepts requests from higher layers in the storage stack, such as VFS, and translates them into equivalent SCSI requests with the help of the middle and lower layers. After the completion of SCSI commands, the upper-level drivers inform the higher layers. The generic SCSI driver, `sg`, allows you to directly send SCSI commands to SCSI devices, bypassing the filesystem layer.

The upper-level SCSI disk drivers are implemented in `/linux/drivers/scsi/sd.c`. The upper-layer SCSI disk drivers self-initialize by calling `register_blkdev` to register as block devices, providing a set of functions through the `scsi_register_driver` function to represent all the SCSI devices.

### Mid layer

The **mid layer** is common to all SCSI operations and contains the core of the SCSI support. The mid layer stitches together the upper and lower layers by defining internal interfaces and providing common services to the upper- and lower-level drivers. It oversees the management of SCSI command queues, ensures efficient error handling, and facilitates power management functions. The upper- and lower-level drivers cannot function without the functionality provided by the mid layer.

The generic mid-level SCSI driver is implemented in `linux/drivers/scsi/scsi.c`. The mid layer abstracts the implementation of the lower-level drivers and transforms commands from the upper layers into equivalent SCSI requests. The mid layer also implements command queuing. When a request is received from the upper layer, the mid layer queues the requests for processing. Once the requests have been served, it receives the response from the lower layer and notifies the upper layer. If a request times out, it is the responsibility of the mid layer to perform error handling or resend the request.

There are a couple of important functions through which the mid layer serves as a bridge between the upper and lower layers, `sd_probe` and `sd_init`. During driver initialization and whenever a new SCSI device is connected to the system, the `sd_probe` function plays a crucial role in determining whether the device is under the management of a SCSI disk driver. If the device falls within the purview of management, `sd_probe` generates a fresh `scsi_disk` structure to serve as its representative entity. When a read or write request is received from higher layers in the storage stack, such as a filesystem, the `sd_init_command` function converts that request into an equivalent SCSI read or write command.

### Lower layer

The **lower-level drivers** are the closest to the hardware. These are the drivers for the various adapters and controllers supported by the operating system. These adapters or controllers are often called **Host Bus Adapters** (**HBAs**). The lower-level drivers provide the actual support for the hardware platform running underneath. The lower-level drivers are vendor-specific and provide an interface to the underlying hardware. For example, `lpfc` is the device driver for Emulex HBAs. The lower-level drivers are present in the `linux/drivers/scsi/` directory.

Now, let's delve deeper into how the SCSI subsystem can be classified as operating within a client-server model.

## The client and server model

The SCSI subsystem receives requests from higher layers in the storage stack to send or retrieve blocks of data from a storage device. When an application initiates a read or write request, the SCSI layer treats this request by transforming it into the equivalent SCSI command. The SCSI subsystem does not handle how data blocks are organized and placed on the storage device; that is the job of the higher layers in the I/O stack. SCSI sends blocks to a destination device, which can either be an individual disk or a **redundant array of independent disks (RAID)** controller.

As the SCSI layer on the operating system side commences an operation on the storage device and the storage device, in turn, responds by performing said operation, this flow of events can be categorized as a client-server exchange model. In SCSI parlance, the two parties are referred to as **initiators** and **targets**. The host operating system that initiates the request is said to act as an SCSI initiator. The destination storage device that receives and processes this request is known as an SCSI target.

The SCSI initiator resides on the host and generates requests on behalf of the higher layers in the I/O stack, such as applications and filesystems. The SCSI target waits for the initiator's commands and then performs the requested data transfers. There has to be an underlying transport mechanism that ensures that the SCSI command from the initiator is delivered to the target. This is implemented through the SCSI transport layer. There are multiple transfer protocols available, such as **Serial Attached SCSI (SAS)** for direct attached disks, and the Fibre Channel or iSCSI for SCSI targets that are part of a **Storage Area Network (SAN)**. The relationship between the SCSI initiator and the target is shown in *Figure 7.4*:

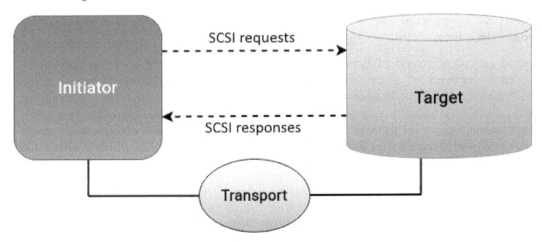

Figure 7.4 – The SCSI initiator and the target

Let us move on to the addressing scheme.

## Device addressing

Linux uses a four-part hierarchical addressing scheme to identify SCSI devices. This combination of four numbers uniquely identifies the location of a SCSI device within a system. If you run `lsscsi` or `sg_map -x` on the command line, you'll see that a sequence of four numbers is used to represent every SCSI device in your system:

```
[root@linuxbox ~]# lsscsi
[0:0:0:0]    disk    ATA      SAMSUNG MZMTE512 400Q  /dev/sda
[4:0:0:0]    disk    ATA      ST9320320AS      SD57  /dev/sdb
[6:0:0:0]    disk    Generic- Multi-Card       1.00  /dev/sdc
[root@linuxbox ~]#
```

This quad addressing scheme is known as **Host, Bus, Target, and LUN** (HBTL), and its fields are explained as follows:

- **Host**: The host represents a controller that can send and receive SCSI commands. The SCSI Host ID is the ID of the HBA, also referred to as the SCSI controller or SCSI adapter. The identifier represents an arbitrary numbering assigned to adapter cards present on the internal system buses. The kernel assigns this number in an ascending manner on the basis of the adapter discovery order. For instance, the first adapter will be assigned zero, the second will be assigned one, and so on.

- **Bus**: This is the bus or channel used within the SCSI controller. A controller can have more than one SCSI bus. This identifier is assigned by the kernel and reflects part of the hardware and firmware architecture of the SCSI controller. Usually, SCSI controllers will only have a single bus. High-end devices such as RAID controllers can have multiple buses.

- **Target**: Each bus can have multiple devices or targets connected to it. The target is the destination device within the bus. This identifier is also assigned by the kernel in the order of target discovery within a given SCSI controller.

- **The Logical Unit Number** (LUN): This is the logical device within the SCSI target, as seen by the host operating system. The LUN is the entity capable of receiving SCSI commands from the host, meaning a disk drive. Each LUN has an exclusive request queue in the kernel's block layer. The LUN identifier is assigned by the storage, making it the only part in the SCSI addressing scheme that is not assigned by the kernel.

The following figure illustrates this addressing mechanism and the path from a host to a SCSI disk. Note that there is a relative SCSI target index associated with a SCSI controller or an HBA. The first SCSI storage target discovered attached to `host0` is assigned the SCSI target (relative index) 0, then 1, and 2, and so on:

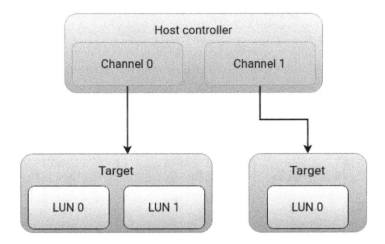

Figure 7.5 – SCSI addressing

If you look inside `/sys/class/scsi_host/`, you will see that hosts 0 to 6 correspond to SCSI controllers:

```
[root@linuxbox ~]# ls -l /sys/class/scsi_host/
total 0
lrwxrwxrwx 1 root root 0 Feb  4 11:12 host0 -> ../../devices/pci0000:00/0000:00:1f.2/ata1/host0/scsi_host/host0
lrwxrwxrwx 1 root root 0 Feb  4 11:12 host1 -> ../../devices/pci0000:00/0000:00:1f.2/ata2/host1/scsi_host/host1
lrwxrwxrwx 1 root root 0 Feb  4 11:12 host2 -> ../../devices/pci0000:00/0000:00:1f.2/ata3/host2/scsi_host/host2
lrwxrwxrwx 1 root root 0 Feb  4 11:12 host3 -> ../../devices/pci0000:00/0000:00:1f.2/ata4/host3/scsi_host/host3
lrwxrwxrwx 1 root root 0 Feb  4 11:12 host4 -> ../../devices/pci0000:00/0000:00:1f.2/ata5/host4/scsi_host/host4
lrwxrwxrwx 1 root root 0 Feb  4 11:12 host5 -> ../../devices/pci0000:00/0000:00:1f.2/ata6/host5/scsi_host/host5
lrwxrwxrwx 1 root root 0 Feb  4 11:12 host6 -> ../../devices/pci0000:00/0000:00:1d.0/usb2/2-1/2-1.6/2-1.6:1.0/host6/
scsi_host/host6
```

Figure 7.6 – The SCSI hosts in Sysfs

Similarly, a structure of the `targetX:Y:Z` format exists in `/sys/bus/scsi/devices/` and is attached to the SCSI bus:

```
lrwxrwxrwx 1 root root 0 Feb  5 07:44 target0:0:0 -> ../../../devices/pci0000:00/0000:00:15.0/0000:03:00.0/host0/tar
get0:0:0
lrwxrwxrwx 1 root root 0 Feb  5 07:44 target0:0:1 -> ../../../devices/pci0000:00/0000:00:15.0/0000:03:00.0/host0/tar
get0:0:1
lrwxrwxrwx 1 root root 0 Feb  5 07:44 target0:0:2 -> ../../../devices/pci0000:00/0000:00:15.0/0000:03:00.0/host0/tar
get0:0:2
lrwxrwxrwx 1 root root 0 Feb  5 07:44 target0:0:3 -> ../../../devices/pci0000:00/0000:00:15.0/0000:03:00.0/host0/tar
get0:0:3
lrwxrwxrwx 1 root root 0 Feb  5 07:44 target3:0:0 -> ../../../devices/pci0000:00/0000:00:11.0/0000:02:03.0/ata3/host
3/target3:0:0
[root@linuxbox ~]#
```

Figure 7.7 – The SCSI targets in Sysfs

The LUNs can be identified through the unique four-level hierarchical addressing scheme, as discussed earlier:

```
[root@linuxbox]# ls -l /sys/bus/scsi/devices/
total 0
lrwxrwxrwx 1 root root 0 Feb  5 07:44 0:0:0:0 -> ../../../devices/pci0000:00/0000:00:15.0/0000:03:00.0/host0/target0
:0:0/0:0:0:0
lrwxrwxrwx 1 root root 0 Feb  5 07:44 0:0:1:0 -> ../../../devices/pci0000:00/0000:00:15.0/0000:03:00.0/host0/target0
:0:1/0:0:1:0
lrwxrwxrwx 1 root root 0 Feb  5 07:44 0:0:2:0 -> ../../../devices/pci0000:00/0000:00:15.0/0000:03:00.0/host0/target0
:0:2/0:0:2:0
lrwxrwxrwx 1 root root 0 Feb  5 07:44 0:0:3:0 -> ../../../devices/pci0000:00/0000:00:15.0/0000:03:00.0/host0/target0
:0:3/0:0:3:0
lrwxrwxrwx 1 root root 0 Feb  5 07:44 3:0:0:0 -> ../../../devices/pci0000:00/0000:00:11.0/0000:02:03.0/ata3/host3/ta
rget3:0:0/3:0:0:0
```

Figure 7.8 – The SCSI LUNs in Sysfs

We'll now explore some major data structures in the kernel that are relevant to the SCSI subsystem.

## Major data structures

The preceding concepts are implemented in the kernel using three major data structures – Scsi_Host, scsi_target, and scsi_device. Of course, these are not the only SCSI-related structures in the kernel. In addition to these three, there are several auxiliary structures, such as scsi_host_template and scsi_transport_template. As the name might suggest, these structures are used to represent some common features for SCSI adapters and transport types. For instance, scsi_host_template provides common content for the host adapters of the same model, including the request queue depth, the SCSI command processing callback function, and the error handling recovery functions. SCSI devices include hard disks, SSDs, optical drives, and so on, and all of these devices have some common functions. These common functions are extracted into templates in the kernel.

These three major structures are as follows:

- Scsi_Host: This is the data structure corresponding to the controller or the HBA, which is located under the SCSI bus. It contains information about the HBA, such as its unique identifier, maximum transfer size, supported features, and host-specific data. Multiple SCSI host structures can exist in the system, each representing a separate host adapter. The SCSI_Host structure acts as the top-level structure for managing SCSI communication.

- scsi_target: This structure corresponds to a target device that is attached to a specific host adapter. It contains information about the target, such as its SCSI ID, **LUN**, and some other flags and parameters. The SCSI target structure is associated with a specific SCSI Host structure and is used to manage communication and commands specific to that target device. The target device can be either physical or virtual.

- scsi_device: This structure represents a LUN within a SCSI target device. It denotes a specific device or partition within a target. When the operating system scans for a logical device connected to the host adapter, it creates a scsi_device structure for the upper-level SCSI driver to communicate with the device. It includes information such as the device's SCSI ID, LUN, and queue depth. It is associated with both a SCSI Target structure and a SCSI Host structure and is used to manage communication and I/O operations for that specific device.

This hierarchical scheme allows the kernel to manage SCSI devices and their communication efficiently. Commands and data transfers can be directed to specific SCSI devices or logical units, and error handling and status tracking can be performed at each level of the hierarchy. The interplay of these structures is highlighted in *Figure 7.9*:

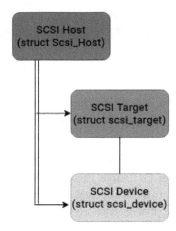

Figure 7.9 – Major SCSI structures

Please note that the actual implementation of SCSI devices in the Linux kernel is far more complex and involves additional data structures and interfaces. However, this simplified diagram demonstrates the basic interconnections between the SCSI Host, SCSI Target, and SCSI Device structures.

### Communicating with SCSI devices

As shown in *Figure 7.10*, there are three different ways of communicating with SCSI devices:

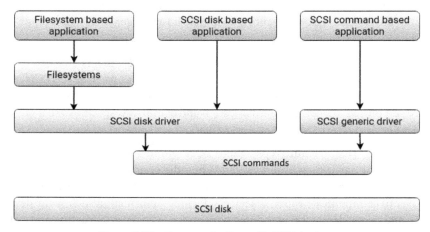

Figure 7.10 – Communicating with SCSI devices

There are explained as follows:

- **Filesystem-based**: The most common method is to access the SCSI device through the interface provided by the filesystem. This is how most regular user-space applications interact with SCSI devices.

- **RAW device access method**: Few applications allow for raw access to SCSI devices. While accessing a device through this approach is less prevalent, utilizing a raw device provides a more direct pathway to the physical device. It grants applications greater control over the timing of input/output operations to the SCSI device. It is important to note that the requests submitted through this approach will go through the block layer. A familiar example of this approach is the dd command in Linux. Using a raw access method does not require address mapping by the filesystem.

- **Pass-through mode**: The pass-through mode allows an application to directly send a SCSI command to the device. The sg3_utils package available in Linux provides a set of utilities that can send SCSI commands to a device via the SCSI pass-through interface provided by the host operating system.

### Interaction between the SCSI and block layers

The SCSI layer and the block layer work together to facilitate the interaction between SCSI devices and the filesystem. The SCSI layer acts as an intermediate layer between the block layer and the device driver specific to the SCSI Host adapter.

The following provides an overview of how the SCSI layer interacts with the block layer:

1. When a filesystem sends an I/O request, such as a read or write operation, it gets translated into a SCSI command by the block layer. The block layer constructs a SCSI **Command Descriptor Block (CDB)** that corresponds to the requested operation and passes it to the SCSI layer.

2. The SCSI mid layer receives the SCSI command from the block layer and performs the necessary processing, including command queuing, error handling, and data transfer.

3. The SCSI mid layer forwards the SCSI command to the appropriate lower-level SCSI device driver associated with the specific SCSI host adapter. The device driver interacts directly with the hardware and sends the SCSI command to the target device over the SCSI bus.

4. Once the SCSI command is executed by the target device, the lower-level SCSI device driver receives the command completion status and communicates this information back to the SCSI mid layer, which then passes it to the block layer.

5. The block layer receives the command completion status from the SCSI mid layer and uses this information to handle any errors, update the I/O request status, and notify the filesystem about the completion or failure of the I/O request.

Please note that this is a summarized view of the interaction between the block and SCSI layers. The block layer provides a standardized interface to the higher layers, such as filesystems. The SCSI layer handles the translation of block-level I/O requests into equivalent SCSI commands, and it manages the communication with the SCSI devices through lower-level device drivers specific to the SCSI host adapter.

## Summary

This chapter focused on two major topics, the device model and the SCSI subsystem in Linux. We started by giving a brief overview of the device model in Linux and how the kernel provides its view in user space through the Sysfs VFS. We then moved on to the exploration of the SCSI subsystem and explained its three-level architecture.

As explained in this chapter, SCSI defines both an interface and a data protocol to connect different types of devices to a system. As a medium, it defines a bus for data transmission, and as a protocol, it defines how devices communicate with each other via the SCSI bus. When an application in user space initiates a write request to store data, the SCSI subsystem converts this write request into a SCSI command, to write the requested data on the specified disk location. It acts as a mediator between the higher layers in the I/O stack and the physical storage. SCSI does not assume responsibility for the assembly of blocks during transport or their physical placement on disk. The side that initiates a request is known as the initiator, while the destination side is known as the target in SCSI terminology. The target of the SCSI protocol can encompass a single physical drive, an HBA, or a RAID controller. The primary duty of the SCSI protocol is to guarantee the successful completion of the write task and report its status to the higher layers.

In the next chapter, we'll discuss the different physical storage options available in today's world, such as mechanical drives, SSDs, and NVMe drives. We'll describe the differences in their design and see how they compare to each other.

# 8
# Illustrating the Layout of Physical Media

*"If I had asked people what they wanted, they would have said faster horses." –*
*Henry Ford*

In the first seven chapters of this book, we explored the organization of storage hierarchy in the Linux kernel, the organization of different layers, the different abstraction methods, and the representation of physical storage devices. We're now done with explaining the software side of things in the storage stack, which means it's time to take a look at the actual hardware and see what lies beneath. I thought it would be best if we got to look at the different types of storage media so that we could not only understand their operations but also see why the Linux kernel uses different schedulers and techniques to handle the different types of drives. In short, getting to know the internals of disk drives will make the information presented earlier in this book a lot more relevant.

The choice of a hardware medium has evolved over the last few years, as there are now considerable options available in the market, not only for enterprise storage but also for personal use. These different storage options are suited to specific environments and workload types. For instance, in some scenarios, people look for capacity-oriented solutions, while in others, maximizing performance is the ultimate concern. Either way, there is a solution available for every scenario. For enterprise environments, vendors offer storage arrays that can implement a hybrid solution and contain a mix of these options.

Most protocols and systems were built and designed with spinning hard drives in mind. The storage stack in Linux is no exception to this rule. When discussing scheduling in the block layer, we saw how techniques such as merging and coalescing are geared toward mechanical drives so that the number of sequential operations can be increased. As this chapter is all about the physical layout and structure of different types of drives, we'll see in detail why spinning drives are slower than the rest.

We'll start by introducing the traditional and oldest form of storage available today, the rotating hard drive. We'll discuss its physical structure, design, and working principles. After that, we'll move on to solid-state drives and see what makes them different from mechanical drives. We'll discuss their internal structure and layout, and explain their operating principles. We'll also briefly discuss the concept of drive endurance and see how it differs for both mechanical and solid-state drives. Finally, we'll discuss the **Non-Volatile Memory Express** (**NVMe**) interface, which has revolutionized the performance of solid-state drives.

We're going to discuss the following main topics:

- Understanding mechanical hard drives
- Explaining the architecture of solid-state drives
- Understanding drive endurance
- Reinventing SSDs with NVMe

# Technical requirements

The material presented in this chapter is operating system-agnostic. As such, there aren't any commands or concepts that are specifically tied to Linux. However, it will help if you have some basic knowledge about the different types of storage media options available today.

# Understanding mechanical hard drives

Mechanical drives, also known as hard disks, magnetic disks, rotating disks, or spinning disks, are the only mechanical component in a modern computer system. We've often addressed them, or, as some might say, *degraded* them, in this book by calling them slower or legacy drives. The truth is that even though the use of mechanical drives has declined in recent years, they are still commonly seen in today's enterprise environments, in a slightly different role. Since there are better storage options available for performance-sensitive applications, hard drives are mostly used for cold data storage. Because of higher capacities and lower costs, mechanical drives are still an integral part of any environment.

Let's briefly describe the major components of a mechanical drive:

- **Platter**: A hard disk consists of multiple thin circular disks, known as platters. All data on a hard drive is recorded on these platters. To maximize capacity, data can be read from and written to both the top and bottom sides of the platter surface. The surface of the platter is magnetized from both ends. The total number of these platters and their storage capacities determine the total capacity of the hard disk.

- **Spindle**: The drive platters rotate under the power of the drive spindle motor, which is designed to maintain constant speeds. The hard disk platter rotates at a rate of several thousand **revolutions per minute (rpm)**, with standard spindle speeds being 5,400 rpm, 7,200 rpm, 10,000 rpm, and 15,000 rpm, as all platters are connected to a single spindle motor. Therefore, they all spin at the same time and rotate at the same speed.

- **R/W (R/W) head**: As a novice, I was under the impression that data is etched on the hard drive in expressed or written form. Well, to burst that bubble, data is expressed by the pattern of a magnetic signal on moving media. Drives have two R/W heads per platter, one each for the top and bottom sides. During data writing, the R/W head modifies the magnetic orientation on the platter's surface, while in data reading mode, the head detects the magnetic orientation on the surface of the platter. It's fascinating to note that the R/W head never touches the surface of the platter.

- **Actuator arms**: The actuator arm assembly is responsible for mounting the **R/W** heads. The actuator arms play a crucial role in the accurate positioning of the R/W heads to the specific locations where data is to be read from or written to.

- **Controller**: The disk controller is a vital component that oversees the operation of the components mentioned earlier, and it interacts with the host system. It carries out instructions from the host, manages the R/W heads, and controls the actuator arms.

Now that we're familiar with the major components of a mechanical drive, let's take a look at the geometry of mechanical drives.

## Looking at the physical layout and addressing

The geometry of a hard drive describes how data is organized on the platters. This organization is based on dividing the platter surface into concentric rings called tracks. A cylinder is a vertical section that intersects the corresponding ring across all platters and is used to refer to specific locations on the disk. A cylinder consists of the same track number on each platter. Each track is further divided into smaller units known as sectors. The sector is the smallest addressable unit on a hard drive. We discussed the concept of block sizes in *Chapter 3*. A block is a group of sectors and is a property of a filesystem. A sector is the physical property of the drive, and its structure is created by the drive manufacturer during initial formatting. Initially, the most common sector size used was 512 bytes. However, some modern drives also use 4 KB sectors. The following is an illustration of the physical arrangement of the mechanical drive:

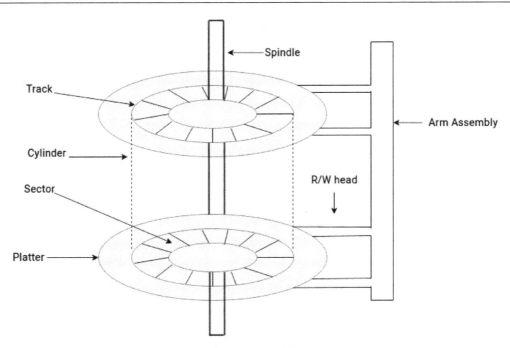

Figure 8.1 – The mechanical drive structure

There are a couple of techniques used to address the physical locations on the hard drive. One such technique is known as **cylinders, heads, and sectors (CHS)**. The physical geometry of a hard disk is usually expressed in terms of CHS. A combination of the CHS numbers can be used to identify any location on the disk. To locate an address on the drive, the host operating system had to be aware of the CHS geometry of the disk.

CHS has now been replaced by **logical block addressing (LBA)**. LBA is another form of disk addressing that simplifies address management on the operating system side. LBA uses a linear addressing scheme to access physical data blocks. When using LBA, instead of addressing sectors through CHS, each sector is assigned a unique logical number. Using it, the hard disk is simply addressed as a single, large device, which simply counts the existing sectors starting at 0. It is then the job of the disk controller to translate LBA addresses into CHS addresses. The host operating system needs to know only the size of the disk drive in terms of the number of logical block addresses.

## Looking at bad sectors

Bad sectors or blocks are areas on the drive that can no longer be written to or read from, either because they have been corrupted or become damaged. In such cases, the drive controller will remap the logical sector to a different physical sector. This can be done transparently, without the knowledge of the host operating system.

There are two different types of bad sectors – hard bad sectors and soft bad sectors. A **hard bad sector** will have suffered physical damage, such as from a physical impact or a manufacturing defect. A hard error is usually uncorrectable, and such a sector cannot be used for further storage of data. A **soft bad sector** is a location on the hard drive that is identified as problematic by the host operating system. Such a sector can be identified by the operating system if the **error-correcting code** (**ECC**) of the sector does not match the information that is written to that location. If an application attempts to retrieve data from a sector and discovers that the ECC does not match the sector's content, this may indicate the presence of a soft sector error. These errors can be rectified and resolved by using various methods.

## Looking at hard drive performance

We've often used the term *seeking* or *seek time* while highlighting the performance limitations of mechanical drives. The seek time of a drive refers to the time taken to position the R/W head across the platter surface, over the correct track. As repeatedly pointed out, random access operations are very costly for mechanical drives. When accessing data on random tracks, the seek time will increase, as the R/W head will have to be continuously moved. The lower the seek time of a drive, the faster the servicing of the I/O requests.

Once the R/W head has been positioned over the correct track, the next task is to position the required sector under the head. To achieve this, the platter is spun to position the requested sector under the R/W head. The total time taken to complete this task is known as rotational latency. This operation is dependent on the speed of the spindle motor. The higher the RPM of the motor, the lower will be the rotational latency. Again, if requests are for adjacent sectors on a track, the rotational latency will be on the lower side. To read and write data on random sectors, the rotational latency will be higher.

The drive heads require alignment over a specific area of the spinning disk to read or write data, resulting in a delay before data can be accessed. To launch a program or load a file, the drive may have to read from various locations, which can lead to multiple delays as the platters need to spin into the correct position each time.

## Understanding where mechanical drives lag

Ever since their inception, it was quite clear that mechanical drives could not possibly match the speeds at which CPUs operate. Response times for mechanical drives are measured in milliseconds, as compared to nanoseconds for CPUs. The presence of mechanical components in the design also limits the performance. It is not that efforts were not made to improve the physical structure of hard drives. For instance, hard drives were equipped with a small on-disk cache to improve performance. Over the years, the speed of the spindle motor has increased from a few hundred rpm to as high as 15,000 rpm. Smaller platter surfaces were also designed for performance improvement. All these factors have contributed to significantly improving the performance of mechanical drives. However, despite all this, even the fastest rotating drives are still far too slow compared to a CPU. A major portion of time is spent on the movement of mechanical parts.

Because of the limitations in the speed of mechanical components, the performance of hard drives falls short in comparison with some modern storage options. The performance of a hard drive is deeply dependent on the read-and-write patterns of the applications. For sequential operations, the performance is significantly better. However, for random access operations, the hard drive performance deteriorates as these operations involve the frequent movement of the R/W head and the continuous rotation of the platter surface. However, despite these drawbacks, mechanical drives are still considered an important part of any enterprise environment. The lower per-gigabyte cost of mechanical drives makes them an excellent choice for cases where capacity is the primary concern.

Now that we have a fundamental understanding of mechanical drives, let's explore solid-state drives and examine how they differ from traditional rotating media.

## Explaining the architecture of solid-state drives

The performance of enterprise storage took a huge leap with the introduction of **SSDs**. SSDs are so-called since they are based on semiconductor materials. Unlike rotating drives, SSDs do not have any mechanical parts and use non-volatile memory chips to store data. Given the absence of moving components, it is no surprise that SSDs are way faster than mechanical drives. They offer a significant upgrade over traditional drives and have gradually replaced mechanical drives as the first-choice storage media.

SSDs make use of flash memory chips for the permanent storage of data. There are two options in this regard, NAND and NOR flash. Most SSDs use NAND flash chips, as they offer faster write and erasure times. At the risk of diving too much into electronics (my least favorite subject in college), a NAND flash is made up of floating-gate transistors, and electrons are stored in a floating gate. When the floating gate contains a charge, it is read as zero. This signifies that data is stored in a cell, contrary to what we typically think (you can see why I didn't like electronics).

The primary components of an SSD are displayed as follows:

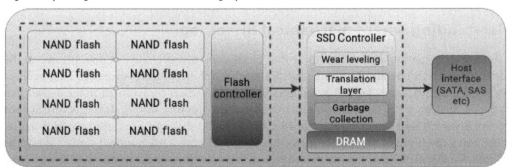

Figure 8.2 – Architecture of an SSD

The SSD controller performs a variety of functions and presents the raw storage in the NAND flash to the host.

## Looking at the physical layout and addressing

Each NAND flash memory in an SSD consists of the following components:

- **Gates**: Floating-gate transistors are a crucial component of SSDs, responsible for the conduction, retention, and release of the electrical charge in the cells, using the electrons stored within the floating gate.

- **Cell**: A cell is a basic unit of storage that can contain a single piece of data, with its electrical charge representing the value of the bit(s). As we will see shortly, cells can hold either a single level or multiple levels of charge.

- **Byte**: A single byte comprises eight cells.

- **Page**: In SSDs, a page is comparable to a sector on a hard drive and represents the smallest unit that can be written to and read from. Typically, a page has a size of 4 KB, although it can be larger than this value.

- **Block**: A block is a collection of pages in an SSD. Erase operations in SSDs are carried out in terms of blocks, which means that all pages within a block must be erased together.

Internally, bits in an SSD are stored in cells, which are then organized into pages. Pages are grouped into blocks, which are in turn encapsulated in a plane, as illustrated in *Figure 8.3*. A die chip typically consists of multiple planes:

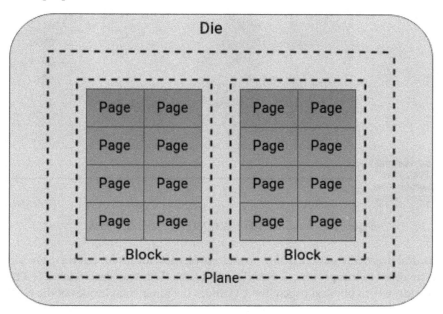

Figure 8.3 – A die layout in SSDs

The terms *SLC* and *MLC* describe the number of bits that are stored per cell. In addition to SLC and MLC, there are also **triple-level cell (TLC)** and **quad-level cell (QLC)** flash drives. Adding more bits per cell of NAND has a significant impact on the performance of a NAND flash. As the number of bits stored per cell increases, the performance of the flash decreases, while the available capacity increases. With SLC NAND, the flash controller only needs to know whether the bit is 0 or 1. With MLC NAND, the cell can have four values – 00, 01, 10, or 11. Similarly, with TLC NAND, the cell can have 8 values, and QLC can have 16 values. The following diagram displays the relationship between various types of SSDs:

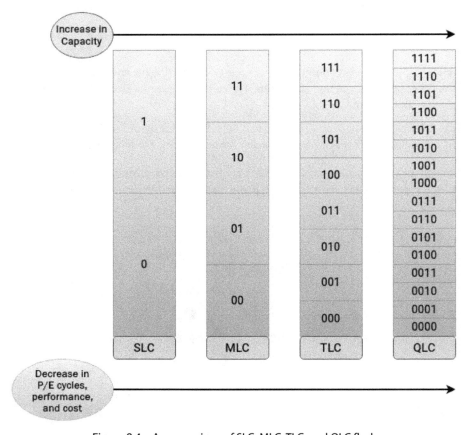

Figure 8.4 – A comparison of SLC, MLC, TLC, and QLC flash

Like a mechanical drive, SSDs also use LBA to address physical locations, but there are some extra components involved in this process. A NAND flash uses a **flash translation layer (FTL)** to map logical block addresses to physical pages. The FTL hides the inner complexities of the NAND flash memory and only exposes an array of logical block addresses to the host. This is deliberately done to emulate a mechanical drive, as most of the stack on the operating system side is optimized to work with mechanical drives.

## Looking at reads and writes

Unlike a hard drive, where a sector is a fundamental unit for all operations, SSDs use different units to perform different operations. The SSDs read and write data at the page level, while all erase operations are performed at the block level. The sectors in mechanical drives can be rewritten repeatedly, whereas pages in SSDs can never be overwritten, and as we'll see in the subsequent section, there is a good reason for that.

The equivalent of sectors in SSDs is pages. It is not possible to read a single cell individually. Read operations align with the native page size of the device. For instance, given a page size of 4 KB, if you want to read 2 KB of data, the flash controller will fetch the full 4 KB page. Similarly, writes also follow the same routine. The write operations align on a page and occur by page size. Given a page size of 4 KB, writing 6 KB of data will use two 4 KB pages.

## Erasing, garbage collection, and the illusion of available space

When an application writes to a NAND flash, the flash must allocate a new blank page for the new data. Erasing NAND flash memory requires a high voltage, and if performed at the page level, it can negatively impact neighboring cells and limit their lifespan. Therefore, SSDs erase data at the block level to mitigate this issue, though it increases the complexity of the erase operation. The erase operation is a crucial factor in determining the lifespan of a NAND flash. The term **Program and Erase Cycle** (**P/E**) reflects the life of an SSD, based on the number of P/E cycles the NAND flash can endure. When a block is written to and erased, that is counted as one cycle. This is important because blocks can be written to a finite number of times, beyond which they cannot write new data anymore.

So, how does an SSD erase data? Let's say that we write some data to an SSD. The write operation, also called the program operation, takes place at the page level. After some time, we realize that the previously written data needs to be updated with some new content. There are two cases here:

- Enough free pages are available in blocks
- Enough free pages are available, but all blocks contain a mix of free, used, and stale pages, or a mix of used and stale pages

Let's say that there are free pages available. The flash controller will write the updated data to any free empty pages, and the older pages will be marked as stale. The pages that are marked as stale are part of a block. It is quite possible that some of the other pages in the same block contain data that is in use. When a page in a block needs to be updated, the flash controller reads the contents of the entire block (which contains the said page) in its memory and computes the updated value of the page. Then, it performs an erase operation on that block. This block erase operation erases the contents of the entire block, including the pages, other than the one that was to be updated. The flash controller then writes the previous contents of the block and the updated value of said page. This entire process is called write amplification. Write amplification refers to a situation where the number of write operations performed by the storage device is more than the number of operations performed by the host device.

As all erase operations are performed at the block level, how can we reclaim space occupied by stale pages? Surely, the controller will not wait for all pages in a block to become stale before erasing them? If that's the case, then the drive will run out of free pages very soon. Clearly, this approach can have some dangerous consequences. This brings us to the second point – how will an SSD cope with a situation when there aren't enough free pages available to accommodate new writes, or all the blocks contain a combination of used and stale pages? To reclaim stale pages, the erase operation needs to be applied at the block level, but where do we put all the pages that are currently in use? The following figure highlights this specific scenario:

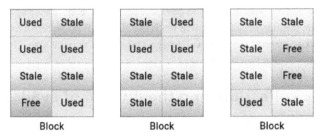

Figure 8.5 – Where do we write incoming data?

That's where the illusion of over-provisioning comes in. There's a lot more space in a flash drive than is visible to the end user. This unallocated space is reserved by the SSD controller for operations such as wear leveling and garbage collection. That extra space comes in handy in situations such as when there is a need to free up stale blocks. The process of cleaning up stale blocks is known as garbage collection. Usually, SSDs can have 20-40% extra capacity than advertised. Flash drive vendors use this trick across the board, from SSDs used for personal systems to SSDs used in enterprise storage arrays. This extra space contributes to improving the endurance and write performance of the SSD.

## Looking at wear leveling

Given the limited number of P/E cycles, the purpose of the wear-leveling operation is to increase an SSD lifespan by making sure that data is distributed evenly across the pages. When data in a particular cell needs to be modified, the wear-leveling process informs the FTL to remap the LBA to point to the new block. Wear-leveling marks the old data as stale. As explained earlier, the current block does not have to be erased. All these decisions are taken to extend the life of a cell. For instance, if a host application frequently updates values in a single cell, and the flash controller modifies the same block again and again, the insulators of this cell will wear out more quickly.

## Looking at bad block management

As each cell can only go through a finite number of P/E cycles, it is important to keep track of cells that have become defective and cannot be programmed or erased anymore. From this point, the cell is considered to be a bad block. The controller keeps a table of all the bad blocks. If pages in the block contain valid data, then existing data in the block is copied over to a new block, and the bad block table is updated.

## Looking at SSD performance

The use of NAND flash in SSDs makes SSDs really fast. Although the performance is still nowhere near as fast as the main memory, it's multiple orders of magnitude faster than a spinning hard drive. The absence of any mechanical components ensures that an SSD is not pinned down by the factors that limit a mechanical drive's performance. Random access operations, which are the Achilles heel of a mechanical drive, are no longer a worry when using SSDs.

## Understanding where SSDs lag

The thing with electronic devices is that they're mostly designed to last for a certain period of time. This is also the case with SSDs; there is a certain life expectancy associated with them. Again, to keep the physics and electronics stuff short and sweet, the write process in SSDs stores electrons, and the erase process drains the voltage in the floating-gate transistor. This sequence of events is known as the P/E cycle. Each NAND cell contains insulators that help to retain voltage in a cell. Every time a cell goes through a P/E cycle, the insulator goes through some damage. The extent of this damage is limited in nature, but over time, this builds up, and eventually, the insulators will lose their capabilities. This may result in voltage leakage, which can result in a change between voltage states. After this, the cell will be considered defective and can no longer be used for storage. If too many cells reach their fate, the drive will cease to work properly. This is why mechanical hard drives have better endurance than SSDs.

You'll often see that the specification sheet for an SSD will contain the number of P/E cycles that it can endure. However, this does not mean that SSDs cannot be used for long-term storage. Although there is a limit associated with their life expectancy, that limit is usually quite long. It can continue to be used for years. There are a lot of tools out there that can check the health and wear level of an SSD. *Table 8.1* highlights some common drive operations for mechanical drives and SSDs, showing their fundamental units:

| Drive | Operation | Unit |
|---|---|---|
| Mechanical disk | Read | Sector |
| | Write | Sector |
| | Update | Sector |
| | Erase | Sector |
| | Bad block management | Sector |
| SSD | Read | Page |
| | Write | Page |
| | Update | Block |
| | Erase | Block |
| | Bad block management | Block |

Table 8.1 – The operational units of SSDs and HDDs

In terms of performance, SSDs offer huge benefits compared to traditional mechanical drives. They have far lower latencies, which has pushed applications toward new performance thresholds. They have now become common not only in enterprise environments but also in personal systems. They do not have any mechanical components, and most drives make use of the NAND flash for persistent storage of data. They have far more complex internal structures and policies than a rotating drive. They are more expensive than mechanical drives but outperform them in almost every other aspect.

Let's briefly touch on the topic of drive endurance before we delve into the world of NVMe drives.

## Understanding drive endurance

There is usually an endurance rating associated with both mechanical drives and SSDs. The endurance of a drive defines multiple things, such as its maximum performance, workload limits, and its **mean time between failures** (**MBTF**). Owing to their contrasting natures, the endurance of mechanical drives and SSDs is measured in differing ways.

The endurance ratings for both types of drives are expressed in different ways. As explained earlier, cells in the NAND flash can go through a finite number of P/E cycles. Once this limit is reached, the cell will become defective. The endurance rating for SSDs is a function of the number of P/E cycles for which the NAND is rated. It is important to note that NAND cells only wear out for write (program) and erase operations. For read operations, this overhead is negligible. The metrics to measure the endurance of SSDs are as follows:

- **Drive Writes per Day** (**DWPD**): The DWPD rating shows how many times you can overwrite an entire SSD each day of its life
- **Terabytes written** (**TBW**): The TBW rating represents how much data can be written to a drive across its entire life, before you may need to replace it

If you have an SSD of 100 GB, with a warranty period of three years and a DWPD rating of 1, that means you can write 100 GB of data to the drive every single day, for the next three years, which means your TBW rating will be 109 TBW:

*100 GB x 365 days x 3 years ≈109 TB*

Mechanical hard drives are different, as they are not impacted by P/E cycles. The magnetic platter surface on mechanical drives supports the overwriting of data. If there is already data in the physical location to be written, the existing data can be directly overwritten with new data. However, while SSD ratings are affected only by write operations and not by the number of read operations, mechanical drives on the other hand are affected by both read and write operations. Hence, the rating for mechanical drives is specified in terms of the number of bytes written and/or read. This workload limit does not have an official term, but going by the terms used for SSD, this can be unofficially called **Drive Writes/Read per Day** (**DWRPD**). The impact of read and write operations on drive endurance is illustrated in *Table 8.2*:

| Drive type | Workload rating | Operation | Effect on endurance |
|---|---|---|---|
| Mechanical disk | DWPD | Read | Decreases |
| | | Write | Decreases |
| SSD | DWRPD | Read | Negligible impact |
| | | Write | Decreases |

Table 8.2 – Drive endurance

*Table 8.2* summarizes the effect of read and write operations on both types of drives. The actual warranted values will differ across the different storage vendors. The workload limits for mechanical drives are also expressed in terms of the TB of data that can be read/written per year. When checking for endurance, keep in mind that terms such as DWPD and TBW are just numbers. It is critical to understand that the warranty period is the key to determining the actual endurance. It's best to use both the warranty period and DWPD when choosing a drive.

Now, let's explore how the NVMe interface has revolutionized traditional SSDs.

## Reinventing SSDs with NVMe

There are several transport protocols that are used to access mechanical and SSDs. Protocols such as SATA, SCSI, and SAS were originally designed for mechanical drives. Hence, these are more geared toward leveraging the potential of rotating drives. With the inception of SSDs, these protocols began to be used for these types of drives as well. Most SSDs, especially in the earlier days, used SATA and SAS ports, just like any other mechanical drive. They would easily fit into existing mechanical drive slots and get connected to the system through a SATA or SAS controller. Despite the major performance gains when using SSDs, the fact that the interfaces, protocols, and command sets that were originally written for mechanical drives were being used for SSDs was considered an overhead, and it was widely thought that this somewhat restricted flash drives from unleashing their full potential.

The NVMe interface was designed specifically for technologies such as the NAND flash. NVMe is often confused as a new type of drive, but technically, it's not. The NVMe is a storage access and transport protocol for SSDs. It acts as a communication interface that operates directly over a PCIe interface. A standard SSD is a drive with SATA or SAS interfaces. These drives are accessed by the host operating system through traditional SCSI protocols. An NVMe SSD uses the M2 physical form factor and uses the NVMe logical interface, developed specifically for these types of drives. The NVMe drive is accessed solely using a PCIe interface. On the host operating system, separate drivers and protocols are used to access NVMe drives. In short, remember the following:

*Every NVMe drive is an SSD, but not every SSD is an NVMe drive.*

NVMe skips the route taken by traditional SSDs and connects directly to the CPU through the PCIe interface, utilizing PCIe slots on the motherboard. The smaller the signal path between storage and CPU, the better the performance. Additionally, PCIe uses four lanes for storage devices, resulting in data exchange that is four times faster than a SATA connection, which only has one lane. When combined with an NVMe SSD, there is an exponential increase in performance.

The performance boost doesn't happen only because of the powerful hardware. The software stack also needs optimization to take full advantage of the hardware. Sometimes, a hardware component can only be as good as the software controlling it. For instance, the SATA and SAS interfaces support a single queue with 32 and 256 commands respectively. On the other hand, NVMe has 64,000 queues and 64,000 commands per queue. That's a difference of a staggering magnitude. There is a separate command set written for the NVMe interface, which is entirely different from all the older SATA and SCSI protocols. The older protocols were designed specifically for mechanical drives and had a large software footprint. NVMe only has a handful of commands, which ensures that a very small number of CPU cycles are spent when processing I/O requests.

As NVMe defines both the communication interface and the method through which storage is presented to the system, this allows for the use of a single driver in the software stack to control the device. For legacy protocols, every vendor is required to develop a driver for every single device to support the required functionality. *Figure 8.6* represents a summarized hierarchy of the storage stack, while highlighting the overhead differences when using the NVMe and SCSI protocols:

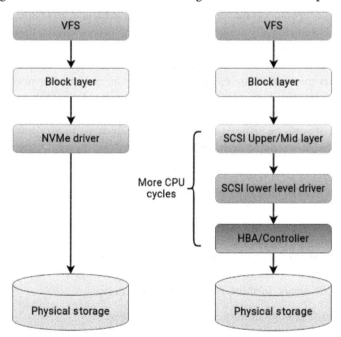

Figure 8.6 – NVMe versus the SCSI stack

When it comes to performance, NVMe drives easily offer the fastest transfer speeds for any available SSD. The read and write performance of NVMe SSDs are far superior to any standard SSD. Due to their exceptional performance, NVMe SSDs are priced significantly higher than standard SSDs, which is not unexpected, since the NVMe interface and protocol were designed to fully utilize the capabilities of SSDs.

## Summary

After spending most of our time examining the software side of things, this chapter focused solely on the actual physical hardware. Because of this, almost all of the information presented in this chapter can be considered platform-agnostic. The hardware capabilities are the same; it's up to the software that drives the hardware to make it reach its full potential.

We discussed the three most common storage options available on the market today – rotating hard drives, SSDs, and NVMe drives. The spinning mechanical drive is one of the oldest forms of storage media available on the market today. It has gone through a few changes over the last few decades, which have improved its performance to some extent. As it consists of several mechanical components, there is a hard limit associated with its performance. After all, the spindle motor that spins its platter surface can only spin so fast. Because of limitations in the hard drive's performance, SSDs came into existence. SSDs do not have any mechanical parts and consist solely of electronic components. They use a NAND flash for permanent storage of data, which makes them extremely fast compared to rotating drives. As the write and erase processes apply and drain voltage from cells, SSDs can endure a limited number of program and erase cycles, which somewhat limits their life span.

Previously, SSDs were limited by the use of protocols and interfaces that were originally designed for mechanical drives. However, with the emergence of NVMe, this limitation was overcome. NVMe was specifically developed for the NAND flash and serves as a storage access and transport protocol for SSDs. Unlike traditional SSDs, NVMe operates directly over a PCIe interface, which makes it significantly faster.

We have now reached the end of *Part 3* of this book. I hope you found the information in it useful. In *Part 4*, we'll discuss and explore some tools and techniques for troubleshooting and analyzing storage performance.

# Part 4:
# Analyzing and Troubleshooting Storage Performance

This part will focus on the different criteria that can be used to evaluate and gauge storage performance. We'll present the different metrics for assessing performance and discuss the different tools and techniques that can be used to investigate performance in each layer of the storage stack. We'll also present some recommended practices that can help to improve storage performance.

This part contains the following chapters:

# Analyzing Physical Storage Performance

*"When you have eliminated the impossible, whatever remains, however improbable, must be the truth."* — *Sir Arthur Conan Doyle*

Now that we're done with understanding the nitty gritty of the storage landscape in Linux, we can put that understanding to practical use. I always like to compare the I/O stack with the OSI model in networking, where each layer has a dedicated function and uses a different data unit for communication. Over the course of the *first eight chapters*, we've increased our understanding of the layered hierarchy of the storage stack and its conceptual model. If you are still following along, you may have gained some understanding of how even the most basic requests from an application have to navigate through numerous layers before being processed by the underlying disks.

Being the good folks that we are, when we work with someone, we can be too willing to be captious and tend to enjoy nitpicking. This leads us to the next phase in our journey – how do we gauge and measure the performance of our storage? There is always going to be a significant performance gap between the compute and storage resources, as a disk is orders of magnitude slower than a processor and memory. This makes performance analysis a very broad and complex domain. How do you determine, how much is too much, and how slow is too slow? A set of values might be perfectly suitable for an environment, while the same set would ring alarm bells elsewhere. Depending upon workloads, these variables differ in every environment.

There are a lot of tools and tracing mechanisms available in Linux that can assist in identifying potential bottlenecks in overall system performance. We're going to keep our focus on the storage subsystem in particular and use these tools to get a sense of what is happening behind the scenes. Some of the tools are available by default in most Linux distributions, which serve as a good starting point.

Here's a summary of what we'll cover in the chapter:

- How do we gauge performance?
- Understanding storage topology

- Analyzing physical storage
- Using disk I/O analysis tools

# Technical requirements

This chapter is a lot more hands-on and requires prior experience with the Linux command line. Most of the readers might already be aware of some of the tools and technologies discussed in this chapter. Having basic system administration skills will be helpful, as these tools deal with resource monitoring and analysis. It would be best to have the required privileges (root or sudo) to run these tools. Depending upon the Linux distribution of your choice, you'll need to install the relevant packages. To install `iostat` and `iotop` on Debian/Ubuntu, use the following:

```
apt install sysstat iotop
```

To install `iostat` and `iotop` on Fedora/Red Hat, use the following:

```
yum install sysstat iotop
```

To install Performance Co-Pilot, you can refer to the installation instructions in their official documentation at the following link: `https://pcp.readthedocs.io/en/latest/HowTos/installation/index.html`.

The usage of these commands is the same on all Linux distributions.

# How do we gauge performance?

There are different lenses through which we can assess the performance of a system. A common approach is to equate the overall system performance with the speed of the processor. If go back to simpler times when single-processor systems were the order of the day and compare them with modern multi-socket, multi-core systems, we'll see that the processor performance has increased by, to put it simply, an epic proportion. If we compare the improvement factor for processor performance with that of a disk, the processor is a runaway winner.

The response times for storage devices are usually measured in milliseconds. For processors and memory, that value is in nanoseconds. This results in a state of incongruity between the application requirements and what the underlying storage can actually deliver. The performance of the storage subsystem has not progressed at the same rate. Therefore, the argument about equating system performance with processor performance has faded away. Just like a chain is only as strong as its weakest link, the overall system performance is also dependent on its slowest component.

Most tools and utilities tend to focus solely on disk performance and do not give much insight into the performance of the higher layers. As we've discovered on this journey, there's a whole plethora of operations happening behind the scenes when an application sends an I/O request to the storage device. Keeping this in mind, we will divide our performance analysis into the following two parts:

- An analysis of physical storage
- An analysis of higher layers in the I/O stack such as filesystems and block layer

For both cases, we're going to explain the relevant metrics and how they can affect performance. An analysis of filesystems and the block layer will be covered in *Chapter 10*. We'll also see how we can check these metrics through the available tools in Linux distributions.

## Understanding storage topology

Most enterprise environments usually contain a mix of the following types of storage.

- **Direct Attached Storage (DAS)**: This is the most common type of storage and is attached directly to a system, such as the hard drive in your laptop. Since data center environments need to have a certain level of redundancy at every layer, the directly attached storage in enterprise servers consists of several disks that are grouped in a RAID configuration to improve performance and data protection.

- **Fibre Channel storage area network**: This is a block-level storage protocol that makes use of fibre channel technology and allows servers to access storage devices. It offers extremely high performance and low response times compared to traditional DAS and is used to run mission-critical applications. It is also far more expensive than other options, as it requires specialized hardware, such as fibre channel adapters, fibre channel switches, and storage arrays.

- **iSCSI SAN**: This is also a block storage protocol that can use the existing network infrastructure and allow hosts to access storage devices. iSCSI SANs utilize the TCP/IP network as a means to transport SCSI packets between the source and target block storage. As it doesn't make use of a dedicated network such as FC SAN, it has a lower performance than FC SAN. However, it is far easier and inexpensive to implement iSCSI SAN, as it doesn't require specialized adapters or switches.

- **Network-Attached Storage (NAS)**: NAS is a file-level storage protocol. Like iSCSI SANs, NAS arrays also rely on the existing network infrastructure and do not require any additional hardware. However, since the storage is accessed through file-level mechanisms, the performance is on the lower side. Nevertheless, NAS arrays are the most inexpensive of the lot and are usually used to store long-term backups.

A simplified comparison of these technologies is shown in *Figure 9.1*. To focus solely on the differences involved in accessing each type of storage, the additional details in the higher layers have been left out:

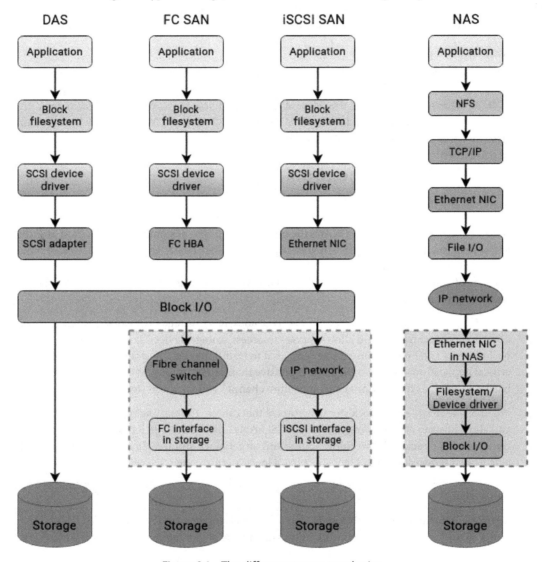

Figure 9.1 – The different storage topologies

We're not going to include fibre switches or any SAN arrays in our discussion. However, keep in mind that there are a lot of components involved in accessing the different types of storage technologies. Every layer warrants careful examination, and as such, you should always have a topology map in mind when diagnosing storage environments.

# Analyzing physical storage

Performance defines how well a disk drive functions when accessing, retrieving, or saving data. There are quite a few yardsticks that can help to define the performance of the disk subsystem. For those of you who have worked with storage vendors while evaluating and purchasing high-end storage arrays, IOPS will be a very familiar term. Vendors like to throw this acronym around a lot and cite a storage system's IOPS as one of its main selling points.

**Input Output Operations per Second (IOPS)** might very well be an entirely useless figure, unless it is coupled with other capabilities of a storage system, such as the response time, the read and write ratio, throughput, and block size. The IOPS figure is usually referred to as *hero numbers*, and it rarely provides any insight into the capabilities of the system unless it is coupled with other metrics. When you purchase a vehicle, you need to know the intricate details, such as its acceleration, fuel economy, and how well it will handle bends and corners. You rarely think about its top speed. Similarly, you need to know all the capabilities of the storage system.

Keeping our focus on the physical disk, we'll first define the **time-based performance metrics**, since they are the ones that explain how and where time is spent. Any time you hear the word **latency** or **delay** while analyzing performance, that usually is an indication of *lost time*. It is the time that could have been spent while working on something, but instead, it was spent waiting for something to happen.

## Understanding disk service time

Let us first develop an understanding of the time-dependent metrics that we need to look for when analyzing physical disks. Once we've gained a conceptual understanding, we'll use specific tools to look for potential bottlenecks. The following figure represents the most common *time-centered* metrics to gauge disk performance:

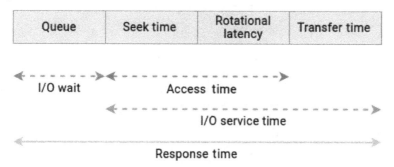

Figure 9.2 – Disk service times

It's important to state here that the aforementioned metrics do not account for the time spent going through the kernel's I/O hierarchy, such as filesystems, the block layer, and scheduling. We're going to take a look at them separately. For now, we're only going to focus on the physical layer.

The terms used in *Figure 9.2* are explained here:

- **I/O wait**: An I/O request can either wait in a queue or be actively served. An I/O request is inserted into the disk's queue before being dispatched for servicing. The amount of time spent waiting in the queue is quantified as I/O wait.

- **I/O service time**: The I/O service time amounts to the time during which the disk controller actively serviced the I/O request. In other words, it is the amount of time an I/O request was not waiting in a queue. The servicing time includes the following:

  - The *disk seek time* is the time taken to move the disk read-write head, with radial movement, to the specified track.

  - Once the read-write head is placed on the correct track, the platter surface rotates to move the exact sector (from where data is to be read from or written to) and line it up with the read-write head. The amount of time spent here is known as *rotational latency*.

  - Once the read-write head is positioned over the correct sector, the actual I/O operation is performed. This amounts to the *transfer time*. Transfer time is the time taken to transfer data to/from the disk from/to the host system.

- **Response time**: The response time or latency is the aggregate of service and wait times and can be thought of as the *round trip time* of an I/O request. It is expressed in milliseconds and is the most consequential term when working with storage devices, as it denotes the entire time from the issuance of an I/O request to its actual completion, as depicted in *Figure 9.3*:

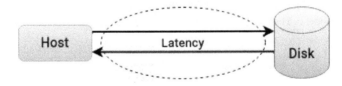

Figure 9.3 – Disk latency

As shown in *Figure 9.4*, storage vendors usually mention the following seek time specifications:

- **Full stroke**: This represents the time taken by the read write head to move from the innermost to the outermost track on the disk

- **Average**: This is the average time taken by the read write head to move from one random track to another

- **Track to track**: This is the time taken by the read write head to move between two adjacent tracks

The disk seek time specifications are shown in *Figure 9.4*:

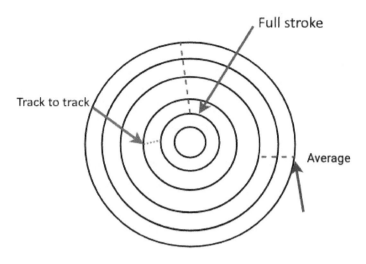

Figure 9.4 – The disk seek time specifications

The transfer rate can be broken down into internal and external transfer rates:

- **Internal transfer rate**: This is the speed at which data is transferred from the disk's platter surface to its internal cache or buffer.

- **External transfer rate**: Once the data has been fetched in the buffer, it is then transferred to the host bus adapter controller via the disk's supported interface or protocol. As highlighted in *Figure 9.5*, the speed at which data is transferred from the buffer to the host bus adapter determines the external transfer rate:

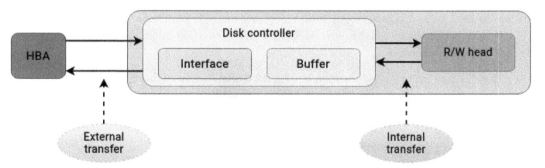

Figure 9.5 – The disk transfer rates

As we explained in *Chapter 8*, unlike mechanical drives, SSDs do not use any mechanical components. Therefore, concepts such as rotational latency and seek time do not apply to them. The *response time* encapsulates all the time-related aspects, and it is the term that is most frequently used when checking for performance-related issues.

## Disk access patterns

The mechanical drives are most affected by the I/O access patterns. The I/O pattern generated by an application can be a combination of sequential and random operations:

- **Sequential I/O**: Sequential I/O operations refer to I/O requests that read from or write data to consecutive or contiguous disk locations. For mechanical drives, this results in a major performance boost, as this requires a very small movement from the read write head. This reduces the disk seek time.

- **Random I/O**: Random I/O requests are performed on non-contiguous locations on the disk, and as you can guess, this results in longer disk seek times, which has a negative impact on disk performance.

Again, the random I/O operations impact the rotating mechanical drives and do not affect the SSDs as such. Although, since reading adjacent bytes on a disk requires a much smaller effort from the controller, sequential operations on SSDs are faster than random operations. However, this difference is much smaller compared to mechanical drives.

## Determining reads/writes ratio and I/O size

IOPS alone do not paint the full picture of the disk's performance and should always be taken with a grain of salt. It is important to look at the size of I/O requests and the ratio of read and write operations. For instance, complex storage systems are designed for specific read-write ratios and I/O sizes, such as `70/30 read write` or `32 KB` block sizes.

Different applications have different requirements and expectations from the underlying drives. It is important to have a rough estimate of the percentage of types of I/O operations that will be performed on a storage device. For instance, online transaction processing applications usually consist of the `70/30 read write` ratio. On the other hand, a logging application might always be busy writing and might require fewer reads.

The size of the I/O request by an application also varies, depending upon the type of the application. In some cases, it is a far more effective approach to transmit larger blocks. The time required to process such a request is longer than a single smaller request. On the other hand, considering the same amount of data, the combined processing and response time of many smaller requests might be greater than a single larger request.

## Disk cache

Modern drives come with an onboard **disk cache** or **buffer**. The disk buffer is the embedded memory in a disk drive that acts as a buffer between the **host bus adapter** (**HBA**) and the disk platter or flash memory that is used for storage.

The following table highlights the effect of cache on different types of I/O patterns:

| I/O type | Read | Write |
|----------|------|-------|
| Random | This is difficult to cache and pre-fetch, as a pattern cannot be predicted. | Caching is extremely effective, as random writes require a lot of disk seek time. |
| Sequential | Caching is extremely effective, as data can be easily pre-fetched. | Caching is effective and can be flushed quickly, as data is to be written to contiguous locations. |

Table 9.1 – The effect of a cache on read/writes

The use of a cache speeds up the process of storing and accessing data from the hard disk. Enterprise storage arrays usually have a huge amount of cache available for this purpose.

## IOPS and throughput

Along with latency, IOPS and throughput define the fundamental characteristics of physical storage:

- **IOPS**: IOPS represent the rate at which I/O operations can take place within a specific time period. The measurement of IOPS will give you the operations per second that the storage system currently delivers.

- **Throughput**: Throughput refers to the volume of data that is transferred from or to the disk drive – in other words, the amount of pizza that you can eat at once. This is also referred to as bandwidth. As throughput measures the actual amount of data transfer, it is expressed in MB or GB per second.

Here are a couple of important things to remember:

- The IOPS figure should always be correlated with latency, read-write ratios, and the I/O request size. When used independently, it does not have much value.

- When processing large amounts of data, the bandwidth statistics might be more relevant than IOPS.

## Queue depth

The **queue depth** dictates the number of I/O requests that can be concurrently handled at one time. In general, this value will not need to be altered. For large-scale SAN environments, in which hosts are connected to storage arrays using Fibre channel HBAs, this becomes a significant value. In that case, there are separate queue depth values for disks, HBAs, and the storage array ports.

If the number of issued I/O requests exceeds the supported queue depth, any new requests will not be entertained by the storage device. Instead, it will return a "queue full" message to the host. Once there is room in the queue, the host will have to resend the failed I/O request. The queue depth settings can impact both mechanical drives and SSDs. Mechanical drives and SSDs that use SATA and SAS interfaces only support a single queue, with 32 and 256 commands. Conversely, NVMe drives have 64,000 queues and 64,000 commands per queue.

In most cases, the default settings for queue depth might be sufficient. Each component in a storage environment has some queue depth settings. For instance, a RAID controller also has its own queue depth, which can be larger than the combined queue depth of the individual disks.

## Determining disk busyness

There are a couple of concepts that determine how much the disk is actually used. They are described as follows:

- **Utilization**: Disk utilization is a fairly common metric that you'll see being reported by various tools. Utilization means that the disk was actively used for a given interval. This value is represented as a percentage of time. For example, a 70% utilization value indicates that if the kernel looked up the disk 100 times, on 70 occasions, it was busy while performing some I/O request. Similarly, a disk that is being 100% utilized means that it constantly serves I/O requests. Again, a fully utilized disk may or may not become a bottleneck. This value needs to be correlated with a few other metrics, such as the associated latency and queue depth. It could be that, although the I/O requests are issued continuously, they're fairly small and sequential; hence, the disk is able to serve them in a timely manner. Similarly, RAID arrays have the ability to handle requests in parallel, and as such, a 100% utilized disk might not be problematic.

- **Saturation**: Saturation means that the amount of requests issued to a disk might be more than what it can actually deliver. This means that we're trying to exceed its rated capacity. When saturation happens, the applications have to wait before being able to read from or write data to disk. Saturation will result in increased response times and impact the overall performance of a system.

## I/O wait

Quite understandably, **I/O wait** is often the most misunderstood metric when checking for performance issues. Although it has an *I/O* in name, I/O wait time is actually a CPU metric, but it doesn't indicate issues with CPU performance. Get it?

I/O wait time is the percentage of time that a CPU was idle, during which the system had pending disk I/O requests. What makes this difficult to comprehend is that it is possible to have a healthy system with a high I/O wait percentage, and it is also possible to have a slow-performing system without a low I/O wait percentage. A high I/O wait means that the CPU is idle while waiting for disk requests to be completed. Let's explain this with a couple of examples:

- For instance, if a process has sent some I/O requests and the underlying disk is unable to immediately fulfill that request, the CPU is said to be in a waiting state, as it is waiting for the request to be completed. Here, the waiting indicates that CPU cycles are wasted and the underlying disk might be slow to respond to I/O requests.

- Then, there's the opposite case. Let's say that process A is extremely CPU-intensive and constantly keeps the CPU busy. Another process running on the system, process B is I/O-intensive and occupies the disk. Even if the disk is slow to respond to requests of process B and becomes a source of a bottleneck for the system, the I/O wait value will be very low in this case. Why? Because the CPU is not idle, as it is wrapped up while serving process A. Therefore, although the I/O wait is on the low side, there could be a potential bottleneck with the storage.

High I/O wait values can be caused by anyone or a combination of the following factors:

- Bottlenecks in the physical storage
- A large queue of I/O requests
- Disks nearing saturation or fully saturated
- Processes in an uninterruptible sleep state, known as the D state (this is fairly common when storage is accessed through **network filesystem (NFS)**)
- Slow network speed in the case of NFS
- High swapping activity

I think we've covered quite a few things to watch out for when doing an analysis of storage devices. Again, if your storage environment contains all the components in a traditional SAN environment, then you need to look for a few more things, such as **fibre channel** (**FC**) switches and any potential bottlenecks on the storage array. To troubleshoot FC switches, you need to establish a basic understanding of the FC protocol.

Let's see how we can identify these red flags using the available tools.

# Using disk I/O analysis tools

We now have developed a basic understanding of what to look for when diagnosing problems with the underlying storage. Most of the time, the problematic behavior is first reported at the application layer, and multiple layers are checked before the actual identification of the issue. The problematic scenario can also be intermittent in nature, which could make it even more difficult to detect. Fortunately, Linux has a broad range of utilities in its toolbox that can be used to identify such problematic behavior. We'll take a look at them one by one and highlight the things of value to look for when troubleshooting performance.

## Establish a baseline using top

`top` is one of the most frequently used commands when troubleshooting performance issues. What makes it so effective is that it can quickly give you the current status of a system and possibly give you a hint about the potential problem. Although most people use it for CPU and memory analysis, there is one particular field that can indicate a problem with underlying storage. As shown in the following output, the `top` command can quickly provide a summarized view of the current system state:

```
top - 19:11:56 up 96 days, 12:38,  0 users,  load average: 9.44, 6.71,
3.75
Tasks: 498 total,   14 running, 484 sleeping,   0 stopped,   0 zombie
%Cpu(s): 20.6%us,  7.9%sy,  0.0%ni, 13.4%id,
57.1%wa,  0.1%hi,  0.9%si,  0.0%st
KiB Mem : 19791910+total, 10557456 free, 80016952 used, 10734470+buff/
cache
KiB Swap:  8388604 total,  5058092 free,  3330512 used. 11555254+avail
Mem
```

As we discussed earlier, a high I/O wait is an indication of a bottleneck at the storage layer. The wa field is the wait average and indicates the potion of time that the CPU had to wait because of the disk. High wait averages mean that the disk does not respond in a timely manner. Although not discussed here, load averages can also increase because of higher wait averages. This is because load averages include disk waiting activity.

The top utility has several options that can provide insight into CPU and memory consumption, but we're not going to focus on them. As our primary concern here is storage, we need to watch out for high values in the wa column and the load averages.

## The iotop utility

The `iotop` command is a `top`-like utility to monitor disk-related activity. The `top` command, by default, sorts output on the basis of CPU usage. Similarly, the `iotop` command sorts processes by the amount of data read and written by each process. It displays columns that highlight the top disk bandwidth consumers in your system. Additionally, it also displays the proportion of time that the thread/process was engaged in swapping and waiting for I/O operations. The I/O priority, both in terms of class and level, is indicated for each process.

It's better to run `iotop` with the `-o` flag, as it will show processes that currently write to disk:

```
Total DISK READ :       231.10 K/s | Total DISK WRITE :      556.40 K/s
Actual DISK READ:       233.13 K/s | Actual DISK WRITE:      593.72 K/s
   TID  PRIO  USER     DISK READ  DISK WRITE  SWAPIN      IO>    COMMAND
23744 be/2 root         0.00 B/s  519.08 K/s  0.00 %   3.03 %  mysql
10395 be/4 root        231.10 K/s  37.32 K/s  0.00 %  1.58 %  java
```

**What to look for**: The `iotop` command shows the amount of data being read from or written to the disk by a process. Check the supported disk read and write speeds, and compare them with the throughput of the top processes. This can also help to identify unusual disk activity by applications and determine whether any process can read or write an abnormal amount of data to underlying disks.

Sometimes, the `iotop` command might complain that delay accounting is not enabled in the kernel. This can be fixed as follows:

```
sysctl kernel.task_delayacct = 1
```

## The iostat utility

The `iostat` command is the most popular tool for disk analysis, as it displays a wide variety of information that can be of help to analyze performance issues. Most of the metrics that were explained earlier, such as disk saturation, utilization, and I/O wait, can be analyzed through `iostat`.

The first line of the output in `iostat`'s disk statistics is a summary since the most recent boot, which shows the mean for the entire time the system has been up. Subsequent lines are displayed as per-second statistics, calculated using the interval specified on the command line, as shown in the following screenshot:

| avg-cpu: | %user | %nice | %system | %iowait | %steal | %idle |  |  |  |  |  |  |  |
|---|---|---|---|---|---|---|---|---|---|---|---|---|---|
|  | 13.28 | 0.00 | 5.92 | 0.20 | 0.00 | 80.60 |  |  |  |  |  |  |  |

| Device: | rrqm/s | wrqm/s | r/s | w/s | rMB/s | wMB/s | avgrq-sz | avgqu-sz | await | r_await | w_await | svctm | %util |
|---|---|---|---|---|---|---|---|---|---|---|---|---|---|
| sda | 0.01 | 0.33 | 0.70 | 2.19 | 0.02 | 0.06 | 55.55 | 0.02 | 6.74 | 4.11 | 7.58 | 0.90 | 0.26 |
| sdb | 0.02 | 0.07 | 37.88 | 79.94 | 1.37 | 0.35 | 29.94 | 0.02 | 0.13 | 1.06 | 0.45 | 0.35 | 4.13 |

Figure 9.6 – The iostat command

This is what to look for:

- The first line, `avg-cpu`, shows the percentage of CPU utilization that occurred while executing in each state.

- The `r/s` and `w/s` numbers give a breakdown of the number of read and write requests issued to the device per second.

- The `avgqu-sz` represents the count of operations that were in either a queued state or actively being serviced. The `await` value corresponds to the average duration between placing a request in a queue and its completion. The `r_await` and `w_await` columns show the average wait time for read and write operations. If you see consistently high values here, the device might be nearing saturation.

- The `%util` column shows the amount of time during which the disk was busy serving at least one I/O request. The utilization value might be misleading if the underlying storage is a RAID-based volume.

The general assumption is that as a device's utilization approaches 100%, it becomes more saturated. This holds true when referring to a storage device that represents a single disk. However, SAN arrays or RAID volumes consist of multiple drives and can serve multiple requests simultaneously. The kernel does not have direct visibility on how the I/O device is designed, which makes this a dubious figure in some cases.

## Performance Co-Pilot

**Performance Co-Pilot (PCP)** is an open source framework and toolkit designed to monitor, analyze, and respond to various aspects of real-time and past system performance data. PCP also includes some utilities to analyze storage system performance. The tools provided by PCP are very similar to the ones included in the `sysstat` package. The PCP tools also include a GUI application to create graphs from the generated metrics and have the ability to save historical data for later viewing. A couple of tools that can assist in storage analysis are as follows:

- `pcp atop`: This provides information similar to both the `iotop` and `atop` commands. The command lists the processes that perform I/O, along with the disk bandwidth they use. Like `ioto` and `top`, `pcp atop` is a good tool for quickly grasping changes occurring on a system.

- `pcp iostat`: The `pcp iostat` command reports live disk I/O statistics, much like the `iostat` command we saw earlier. As shown in *Figure 9.7*, the columns in the output are similar to that of `iostat`:

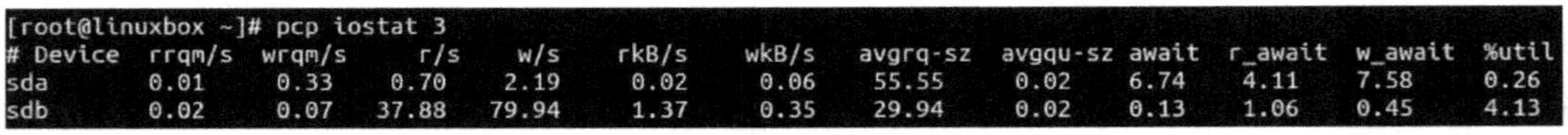

```
[root@linuxbox ~]# pcp iostat 3
# Device  rrqm/s  wrqm/s    r/s    w/s   rkB/s  wkB/s  avgrq-sz  avgqu-sz  await  r_await  w_await  %util
sda         0.01    0.33    0.70   2.19   0.02   0.06    55.55     0.02     6.74    4.11     7.58    0.26
sdb         0.02    0.07   37.88  79.94   1.37   0.35    29.94     0.02     0.13    1.06     0.45    4.13
```

Figure 9.7 – pcp iostat

When troubleshooting disk performance or resource-related issues, `vmstat` can provide valuable information, as it can help to identify disk I/O congestion, excessive paging, or swapping activity.

## The vmstat command

The vmstat command, derived from "virtual memory statistics," is a native monitoring utility included in nearly all Linux distributions. As shown in *Figure 9.8*, it reports information about processes, memory, paging, disks, and processor activity:

```
procs -----------memory---------- ---swap-- -----io---- -system-- ------cpu-----
 r  b   swpd   free    buff  cache   si   so    bi    bo   in   cs  us sy id wa st
 3  0 3240532 2163360 10468 115039776 0    1   197    72    0    0   10  6 84  0  0
 2  0 3240532 2100884 10468 115045008 0    0  2103    96 19047 24440 10  5 84  1  0
 2  1 3240532 2095672 10468 115050480 0    0  2179    26 18153 22752 12  5 82  1  0
 2  0 3240532 2148732 10468 115054800 3    0  2269     9 21943 24114 26  6 67  1  0
```

Figure 9.8 – The vmstat command

**What to look for**: The b column in the output shows the number of processes blocked while waiting for a resource, such as disk I/O. Additional information most useful for troubleshooting I/O issues is as follows:

- si: This field represents the amount of memory, in kilobytes, that is swapped in from the swap space on the disk to the system's memory per second. A higher value in the si field indicates increased swapping activity, suggesting that the system frequently retrieves data from the swap space.

- so: This field represents the amount of memory, in kilobytes, that is swapped out from the system's memory to the swap space on disk per second. A higher value in the so field indicates increased swapping activity, which can occur when the system is under memory pressure and needs to free up physical memory.

- bi: This field specifically refers to the data transfer rate from disk to memory. A higher value in the bi field indicates increased disk read activity.

- bo: This reflects the output activity or the amount of data written from the system's memory to the disk. A higher value in the bo field suggests increased disk write activity, indicating that data is written from memory to the disk frequently.

- wa: This field represents the percentage of time that the CPU is idle while the system waits for I/O operations to complete. A higher value in the wa field suggests that the system experiences I/O bottlenecks or delays, with the CPU frequently waiting for I/O operations to complete.

## Pressure Stall Index

**Pressure Stall Index (PSI)** is a relatively new addition to the troubleshooting toolset in Linux and offers a new way to obtain utilization metrics for memory, CPU, and disk I/O. Latency spikes can occur when there is contention for CPU, memory, or I/O devices, resulting in increased waiting times for workloads. The PSI feature identifies this and prints a summarized view of this information in real time.

The PSI values are accessed through the /proc pseudo filesystem. The raw global PSI values appear in the /proc/pressure directory in files called cpu, io, and memory. Let's take a look at the io file:

```
[root@linuxbox ~]# cat /proc/pressure/io
some avg10=51.30 avg60=41.28 avg300=23.33 total=84845633
full avg10=48.28 avg60=39.22 avg300=22.78 total=75033948
```

**What to look for**: The avg fields represent the *percentage* of time in the last 10, 60, and 300 seconds, respectively, that processes were starved of disk I/O. The line prefixed with some represents the portion of time during which one or more tasks were delayed due to insufficient resources. The line prefixed with full represents the percentage of time when all tasks were delayed due to resource constraints, indicating the extent of unproductive time. This is a bit similar to the load averages in top. The output here shows high values for the 10-, 60- and 300-second interval averages, which indicates that processes are being stalled.

To summarize, Linux offers a plethora of utilities to analyze the performance of your system. The tools that we covered here are not only used for storage analysis but also to establish an overall picture of a system, including its processor and memory subsystem. Every tool offers a wide range of options that can be used if we want to analyze a particular aspect. We highlighted the major indicators to look for when using each of these tools, but since every environment consists of different variables, there is no *fixed* approach to troubleshooting.

## Summary

Troubleshooting performance issues is a complex matter, as it can take a long time to diagnose and analyze them. Of the three major components in an environment – storage, compute, and memory – storage is the slowest. There is always going to be a mismatch in their performance, and any degradation in disk performance can impact the overall operation of the system.

Keeping this objective in mind, we divided this chapter into two sections. In the first section, we explained the most important metrics that you should understand before troubleshooting any issues. We discussed the time-related metrics related to storage devices, CPU wait averages, disk saturation, and disk utilization, and the different access patterns when reading from or writing to physical disks.

In the second part, we saw the different ways in which we can analyze the metrics highlighted in the first section. There are a lot of mechanisms available in Linux that can assist to identify potential bottlenecks in overall system performance. We used tools such as top, PSI, iostat, iotop, and vmstat to analyze disk performance.

In the next chapter, we will continue our analysis of the storage stack and focus on the higher layers, such as the block layer and filesystems. For this purpose, we'll make use of the different tracing mechanisms available in Linux.

# 10

# Analyzing Filesystems and the Block Layer

Read or write access to storage devices usually happens after passing through several intermediary layers, such as filesystems and the block layer. There is also the page cache, where requested data is preserved before being lazily committed to the underlying storage. So far, we've tried to understand the different factors that can affect disk performance and examined the important metrics associated with physical disks, but, as Sherlock Holmes would say, "*Perfectly sound analysis, but I was hoping you'd go deeper.*"

Applications tend to interact with the filesystem, not with the physical storage. It is the job of a filesystem to translate the application's request and send it down to the lower layers for further processing. The request will go through further processing in the block layer and be eventually scheduled for dispatch to the storage device. Each stage in this hierarchy will add its own processing overhead. Therefore, it is extremely important to examine the behavior of the filesystem and block layer to perform any performance analysis.

In this chapter, we will focus on the techniques that can be used to investigate the filesystem and block layer. At this stage, I would like to think that the first six chapters helped us to build a decent understanding of these layers (*I certainly hope so*). Becoming acquainted with the relevant analysis methodologies should not be a problem.

Here's a summary of what we'll be covering:

- Investigating filesystems and the block layer
- The different types of filesystem I/O
- What causes filesystem latency?
- Identifying the target layers
- Finding the right tools

# Technical requirements

This chapter will focus on the **BPF Compiler Collection (BCC)** performance tools in Linux. Having basic system administration skills will be helpful, as these tools deal with system-level monitoring and analysis. Understanding processes, system resource utilization, and performance metrics will assist you to interpret the results obtained from the BCC tools. It would be best to have the required privileges (`root` or `sudo`) to run these tools.

The operating system packages relevant to this chapter can be installed as follows:

- For Ubuntu/Debian:

  - `sudo apt install strace`

  - `sudo apt install bpfcc-tools`

- For Fedora/CentOS/Red Hat-based systems:

  - `sudo yum install strace`

  - `sudo yum install bcc-tools`

# Investigating filesystems and the block layer

Given that storage is a lot more sluggish than other components in a system, it is no surprise that, very often, performance issues are related to I/O. However, simply categorizing a performance issue as I/O-based is an oversimplification.

Filesystems are the first point of contact for an application and are considered to be sandwiched between the application and physical storage. Traditionally, physical storage has always been the center of attention while doing any performance analysis. Most tools focus on the utilization, throughput, and latency of the physical drives, while leaving out the other aspects of an I/O request. Scrutinizing storage usually begins and ends with physical disks, making filesystems analysis an oversight.

Similarly, the events happening in the block layer also tend to slip under the radar when it comes to performance analysis. The tools that we discussed in *Chapter 9* usually provide averaged-out values over a specific interval, which can often be misleading. For instance, let's say an application generates the following number of I/O requests in a 10-second interval:

| Second | Number of requests | Second | Number of requests |
|--------|--------------------|--------|--------------------|
| 1 | 10 | 6 | 20 |
| 2 | 15 | 7 | 5 |
| 3 | 500 | 8 | 15 |
| 4 | 20 | 9 | 8 |
| 5 | 5 | 10 | 2 |

Table 10.1 – The averaged out stats for I/O requests

If I collect I/O statistics after every 10 seconds, then the number of average I/O requests issued per second will be 60 – that is, the total number of requests divided by the interval. The mean value might be considered normal, but it completely ignores the burst of I/O requests issued around the three-second mark. The tools that provide disk-level statistics do not provide any insight on a per-I/O basis.

The conventional approach has always been to gather information from the bottom end of the filesystem – that is, physical disks. However, this a multifaceted problem and involves analyzing the following layers:

- **System and library calls**: Applications use the generic system call interface to request resources from the kernel space. When an application calls a function that is provided by the kernel, then the execution time is spent inside the kernel space. This function is known as a **system call**. Library calls, conversely, are executed in user space. When an application wants to utilize functions defined in a programming library, such as the GNU C-library, it sends a request known as a **library call**. To accurately assess performance, it's essential to measure the time spent in both the kernel and user space. By tracing these calls, it's possible to gain valuable insights into how the application behaves and identify any potential issues, such as resource contention or locking that may cause a process to become stuck.

- **VFS**: As we explained throughout this book, VFS acts as the interface between the user and the backing filesystem. It decouples the application's file operations from the specific filesystem, masking the implementation details behind generic system calls. The VFS also includes the page cache, inode, and dentry cache to speed up disk access. Analyzing VFS can prove helpful for general workload characterization, to identify an application's operational patterns over time, and to pinpoint how the application uses the different types of available cache.

- **Filesystems**: Every filesystem uses a different approach to organizing data on disk. As we explained in *Chapter 9*, it is important to characterize the type of workload that the filesystem will be managing – for instance, access patterns of an application, synchronous and asynchronous operations, the ratio of read and write requests, the cache hit and miss ratio, and the size of I/O requests. Internally, filesystems perform operations such as read-ahead, pre-fetching, locking, and journaling, which can affect overall I/O performance in one way or another.

- **Block layer:** When an I/O request enters a block layer, it can be mapped onto another device, such as LVM, software **Redundant Array of Independent Disks (RAID)**, or a multi-pathed device. It is commonplace to have a filesystem created on top of these logical devices. In such cases, with any filesystem I/O, the corresponding tasks for these techniques require resources that may be the source of an I/O contention, such as RAID striping or multi-pathing I/O drivers.

- **Scheduler:** The choice of a disk scheduler can also impact the I/O performance of an application. A scheduler can use techniques such as merging and sorting, which can change the eventual order in which a request lands on the disk. As we learned in *Chapter 6*, the Linux kernel offers different flavors of disk schedulers. Some I/O schedulers are only suited for high-end storage devices, while others work well with slower drives. As each environment is different, multiple factors need to be taken into account before deciding the appropriate disk scheduler.

- **Physical storage:** The physical layer is usually the point of focus in any troubleshooting scenario. We covered the part about analyzing the different physical disk metrics in *Chapter 9*.

Although not covered here, it's important to know that it is possible to bypass the filesystem and write data directly to the physical storage. This is known as **raw access**, and a device accessed through such methods is known as a **raw device**. Some applications, such as databases, are capable of writing to raw devices. The primary reason for this approach is that any layer of abstraction, such as a filesystem or a volume manager, adds processing overhead. Filesystems make use of a buffer cache to cache read and write operations, deferring their commitment to the disk until later. With the absence of a filesystem, large applications such as databases are able to bypass the filesystem cache, which allows them to manage their own cache. This approach provides more granular control over device I/O and may aid in testing the raw speed of storage devices, as it bypasses any additional processing overhead.

*Figure 10.1* highlights some factors that can affect the I/O performance of an application:

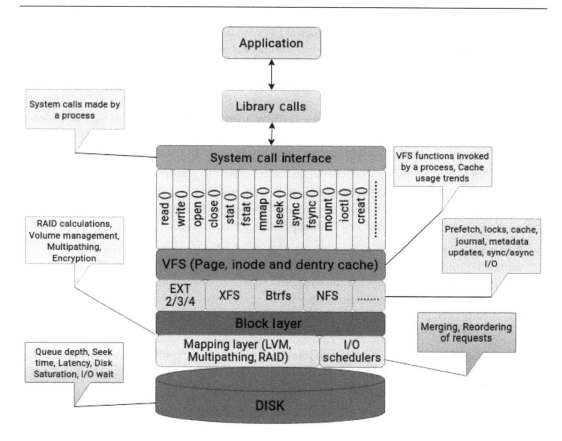

Figure 10.1 – The factors affecting an application's I/O performance

In summary, the different layers in the I/O stack can influence an application's I/O performance in various ways. Therefore, when troubleshooting any performance problem, breaking it down into smaller pieces is the first step; simplify the problem by removing as many layers as possible.

## The different types of filesystem I/O

There are too many different types of I/O requests that can be issued to a filesystem. For the sake of clarity, we'll consider an I/O request issued by a process as logical I/O, while the actual operation that was performed on the disk will be called physical I/O. As you can probably guess, the two are not equal. **Logical I/O** refers to the process of reading or writing data at the logical level, meaning at the level of the filesystem or application. Conversely, **physical I/O** involves the transfer of data between the storage device and memory. It is during this stage that the data is moved at the hardware level and managed by a hardware device such as a disk controller.

Disk I/O can be inflated or deflated. A single logical I/O request may result in multiple physical disk operations. Conversely, a logical request from a process may not require any physical I/O from the disk.

To elaborate on the concept, let's take a look at some of the factors that make the two types of requests disproportionate:

- **Caching**: The Linux kernel heavily uses available memory to cache data. If data is loaded from the disk, it is kept in the cache so that any subsequent access to the same data can be readily served. If a read request by an application is served from the cache, it will not result in a physical operation.

- **Writeback**: As filesystem writes are cached by default, this also contributes to the difference in the number of physical and logical operations. The writeback caching mechanism defers and coalesces write operations before eventually flushing them to disks.

- **Prefetching**: Most filesystems have a pre-fetching mechanism through which they can prefetch sequentially adjacent blocks in the cache while a block is read from the disk. The filesystem anticipates the data that an application will need and reads it into memory before the application actually requests it. The pre-fetching operations make sequential reads very fast. If the data has already been pre-fetched in the cache, the filesystem can avoid future trips to the physical storage, thereby reducing the number of physical operations.

- **Journaling**: Depending upon the type of journaling technique being employed by the filesystem, the number of write operations can be doubled. At first, they will be written to the filesystem journal, and then flushed to disk.

- **Metadata**: Every time a file is accessed or modified, the filesystem will need to update its timestamps. Similarly, when writing any new data, the filesystem will also need to update its internal metadata, such as the number of used and free blocks. All these changes require physical operations to be performed on disk.

- **RAID**: This can be often overlooked, but the type of RAID configuration on the underlying storage can have a huge say in determining whether additional writes are necessary. For instance, operations such as striping data across multiple disks, writing parity information, creating mirrored copies, and rebuilding data will incur additional writes.

- **Scheduling**: I/O schedulers usually employ techniques such as merging and reordering to minimize disk seeks and improve disk performance. Hence, multiple requests can be consolidated into a single request in the scheduling layer.

- **Data reduction**: If any compression or deduplication is performed, the amount of physical I/O requests performed on disks will be lower than the logical requests initiated by an application.

# What causes filesystem latency?

Latency, as we discussed in *Chapter 9*, is the single most important metric in any performance measurement and analysis. From the filesystem's perspective, latency is measured as the time from which a logical request was initiated to the time it was completed on the physical disk.

The latency endured because of the bottlenecks in physical storage is one factor that adds to overall filesystem response time. However, to reiterate our discussion from the previous section, as filesystems do not simply hand over an I/O request to the physical disk, latency can be experienced in more than one way, such as the following:

- **Resource contention**: If multiple processes concurrently write to a single file, then this can impact filesystem performance. File locking can be a significant performance issue for large applications, such as databases. The purpose of locking is to serialize access to files. Filesystems in Linux use the generic VFS methods for locking.

- **Cache misses**: The purpose of caching data in memory is to avoid frequent trips to disks. If an application is configured to avoid using the page cache, then it can experience some delays.

- **Block size**: Most storage systems are designed to work with a specific block size, such as 8 K, 32 K, or 64 K. If the issued I/O requests are of large sizes, they will first need to be broken down into suitable sizes, which will involve extra processing.

- **Metadata updates**: Filesystem metadata updates can be a major source of latency. Updating filesystem metadata involves performing several disk operations, including seeking the appropriate disk location, writing the updated data, and then synchronizing the disk cache with the disk. Depending on the size and location of the metadata being updated, this sequence can take a significant amount of time, especially if the filesystem is heavily used and the disk is busy with other operations. This may result in a backlog of requests and an overall slowdown in filesystem performance.

- **Breakdown of logical I/O**: As explained earlier in the previous section, a logical I/O operation may need to be broken down into multiple physical I/O operations. This may increase the filesystem latency, as each physical I/O operation requires additional disk access time, which will result in additional processing overhead.

- **Data alignment**: File system partitions must be correctly aligned with the physical disk geometry. Incorrect partition alignment will cause reduced performance, especially with regard to RAID volumes.

Given the plethora of things that can affect the I/O performance of an application, it is no surprise that most people are reluctant to explore this avenue and merely focus on disk-level statistics, which are far easier to understand. We've so far only covered some common issues that can impact the life of an I/O request. Troubleshooting is a complex skill to master, and it can be a difficult decision to determine a good starting point. Adding to the confusion is the wealth of tools that can be used for performance analysis. Even though we're only focusing on the storage side of things here, it is impossible to cover the long list of tools that can help us in our goal in one way or another.

# Identifying the target layers

The following table summarizes the different target layers for performance analysis and presents the pros and cons of each approach:

| Layer | Pros | Cons |
|---|---|---|
| Application | Application logs, specific tools, or debugging techniques can determine the scope of the problem, which can aid in subsequent steps. | Debugging techniques are not common and vary for each application. |
| System call interface | It's easy to trace the calls generated by a process. | It's difficult to filter, as there are multiple system calls for the same function. |
| VFS | Generic calls are used for all filesystems. | There is a need to isolate the filesystem in question, as tracing may include data for all filesystems, including pseudo filesystems. |
| Filesystems | Filesystems are the first point of contact for an application, which makes them an ideal candidate for analysis. | There are very few filesystem-specific tracing mechanisms available. |
| Block layer | Multiple tracing mechanisms are available, which can be used to identify how requests are handled. | Some components, such as schedulers, do not offer a lot of tunables. |
| Disk | This is easier to analyze, as this does not require a deep understanding of higher layers. | This does not paint a clear picture of an application's behavior. |

Table 10.2 – Comparing the pros and cons of analyzing each layer

The general consensus (and it definitely has some merit) is that investigating each layer is way too laborious! Enterprises that have dedicated performance analysis engineers make it a habit to go through every tiny detail and identify the potential bottlenecks in a system. However, the more common approach in recent times has been to add more compute power, especially for cloud-based workloads. Adding more hardware resources to an application, as it becomes resource-hungry, is the new normal. Troubleshooting performance issues is often skipped in favor of migrating applications to better hardware platforms.

# Finding the right tools

Trying to dig deep into an application's behavior can be a daunting task. The abstraction layers in the I/O stack do not make our job easier in this regard. To analyze each layer in the I/O hierarchy, you must have a decent grasp of the concepts used in each layer. The job is made even tougher when you include the application in this setup. Although the tracing mechanisms in Linux can help to understand the patterns generated by an application, it is not possible for everyone to have the same level of visibility about the design and implementation details of the application.

If you're running a critical application, such as an **Online Transaction Processing (OLTP)** database that processes millions of transactions every day, it can be helpful to know where CPU cycles are wasted. For instance, there are several service-level agreements associated with a transaction, and it has to be completed within a few seconds. If a single transaction is required to be completed within 10 seconds, and only one second is spent processing the filesystem and disk I/O, then clearly your storage is not a bottleneck, as only 10 percent of the total time is being spent in the I/O stack. If the application was blocked at the filesystem level for five seconds, then clearly some tweaking is required.

Let's take a look at the available options that we have to analyze the I/O stack. Note that this is not a complete list of tools by any means. The BCC itself contains an abundance of such tools. The tools presented as follows have just been cherry-picked based on personal experience.

## Tracing application calls

The **strace utility** is one of the most well-known tools in Linux, which displays information about system calls made by a process. The `strace` command helps identify the kernel function on which a program spends its time. For instance, the following command provides a summarized report and shows the frequency and time spent on each system call. The `-c` switch displays the count. Here, `myapp` is just a simple user space program:

```
[root@linuxbox ~]# strace -c ./myapp
% time     seconds  usecs/call     calls    errors syscall
------ ----------- ----------- --------- --------- ----------------
 46.32    0.000359         359         1           execve
 14.19    0.000110          36         3           openat
  7.74    0.000060          20         3           close
  6.84    0.000053          10         5           mmap
  6.58    0.000051          12         4           mprotect
  3.35    0.000026           6         4           brk
  3.23    0.000025          25         1           munmap
  2.97    0.000023           7         3           fstat
  2.45    0.000019          19         1           write
  2.06    0.000016           8         2           read
  2.06    0.000016           8         2         1 arch_prctl
  1.42    0.000011          11         1         1 access
  0.77    0.000006           6         1           lseek
------ ----------- ----------- --------- --------- ----------------
100.00    0.000775          24        31         2 total
```

Figure 10.2 – Tracing system calls using strace

This command can prove useful to pinpoint some types of process performance bottlenecks. To filter the output and only show stats for a specific system call, use the -e flag:

```
[root@linuxbox ~]# strace -e execve -c ./app
% time     seconds  usecs/call     calls    errors syscall
------ ----------- ----------- --------- --------- ----------------
100.00    0.000468         468         1           execve
------ ----------- ----------- --------- --------- ----------------
100.00    0.000468         468         1           total
[root@linuxbox ~]#
```

Figure 10.3 – Filtering specific calls

Let's take it up a notch and see whether we can make something out of the actual trace output. You can also print the timestamps and the time spent on each system call. The trace output can be saved to a file using the -o flag:

```
strace -T -ttt -o output.txt ./myapp
```

Focusing only on the subset that corresponds to the I/O portion of the application, note the number after the equal sign in the first write call. We can see that the write call was able to buffer all data into a single write function call. The application wrote 319,488 bytes in 156 microseconds:

```
1678639539.375220 execve("./myapp", ["./myapp"], 0x7ffebe7f6f90 /* 34 vars */) = 0 <0.000568>
1678639539.391952 brk(NULL)                = 0x115a000 <0.000025>
1678639539.392033 arch_prctl(0x3001 /* ARCH_??? */, 0x7ffe1fe05af0) = -1 EINVAL (Invalid argument) <0.000024>
1678639539.410730 access("/etc/ld.so.preload", R_OK) = -1 ENOENT (No such file or directory) <0.000034>
1678639539.410840 openat(AT_FDCWD, "/etc/ld.so.cache", O_RDONLY|O_CLOEXEC) = 3 <0.000028>
[.............]
1678639539.413325 fstat(3, {st_mode=S_IFREG|0644, st_size=0, ...}) = 0 <0.000025>
1678639539.413406 write(3, "\1\0\0\3\0\0\0\3\0\0\0\0\0\0\2\0\0\0\4\0\0\0\4\0\0\0\0\0 A"..., 319488) = 319488 <0.
000156>
1678639539.413611 close(3)                = 0 <0.000289>
```

Figure 10.4 – Analyzing the strace output

The strace command can also be attached to a running process. The strace output is quite substantial, and you often have to laboriously look through a great deal of information before you get somewhere. This is why it is a good idea to know about the most frequently generated calls by an application. For I/O analysis, focus on common system calls, such as open, read, and write. This can help in understanding the I/O pattern of an application from the application's perspective. Although strace doesn't tell you what the operating system did with the I/O requests afterward, it does tell you what the application generates.

To summarize, for a quick analysis, do the following:

- Generate a summary of the system calls being generated by the application.

- Check the execution times of each call.

- Isolate the calls you want information about. For I/O analysis, focus on read and write calls.

## Tracing VFS calls

At the very beginning of your investigation, analyzing VFS can be beneficial for general workload characterization. It can also be helpful to identify how efficiently an application makes use of the different types of available caches in VFS. The BCC program contains tools such as vfsstat and vfscount, which can help to understand the events in VFS.

The vfsstat tool shows a statistics summary for some common VFS calls, such as read, write, open, create, and fsync:

```
[root@linuxbox ~]# vfsstat
TIME        READ/s  WRITE/s  FSYNC/s   OPEN/s CREATE/s
13:20:57:     8739     7703        0       15        0
13:20:58:     8070     8046        0       10        0
13:20:59:     9065     8848        0      153        0
13:21:00:     8709     8680        0       13        0
13:21:01:     8287     8264        0       10        0
13:21:02:     7996     7951        0       16        0
13:21:03:     8443     8417        0       12        0
13:21:04:     8909     8882        0       12        0
13:21:05:     7964     7934        0       14        0
13:21:06:     8709     8687        0       10        0
```

Figure 10.5 – The vfsstat output

In addition to the **READ** and **WRITE** calls, keep an eye out for the **OPEN** column. This shows the number of files opened per second. A sudden increase in the number of open files can greatly increase the number of I/O requests, especially for metadata operations.

Running these tools alone might not offer much insight. A good use of these is to run them in conjunction with some disk analysis tools, such as iostat. This will allow you to compare logical I/O requests with physical I/O requests.

One limitation with vfsstat is that it doesn't segregate the I/O activity at the filesystem level. Another program, fsrwstat, traces the read and write functions and breaks them down for the different available filesystems. The following figure shows the breakdown of the number of read and write calls for the different filesystems:

```
[root@linuxbox ~]# fsrwstat
Attaching 7 probes...
Tracing VFS reads and writes... Hit Ctrl-C to end.
14:06:14
@[devpts, vfs_write]: 1
@[pipefs, vfs_write]: 1
@[devtmpfs, vfs_read]: 1
@[anon_inodefs, vfs_read]: 1
@[sockfs, vfs_read]: 5
@[sockfs, vfs_write]: 6
@[proc, vfs_read]: 13
@[pipefs, vfs_read]: 15
@[xfs, vfs_write]: 11937
@[xfs, vfs_read]: 11937
```

Figure 10.6 – The fsrwstat output

Continuing with the output of vfsstat, if you notice a large number of files are open, consider using filetop. This shows the most frequently accessed files on your system and displays their read and write activity:

```
14:34:02 loadavg: 1.68 1.54 0.82 1/350 175981

TID     COMM            READS   WRITES  R_Kb    W_Kb    T FILE
175977  cp              2182    0       279296  0       R tfile2
175977  cp              0       2181    0       279168  R tfile145
175969  cp              0       214     0       27392   R tfile144
175969  cp              214     0       27392   0       R tfile2
175959  cp              0       214     0       27392   R tfile1
175959  cp              214     0       27392   0       R tfile2
175956  cp              128     0       16384   0       R tfile2
175956  cp              0       128     0       16384   R tfile143
```

Figure 10.7 – The filetop output

The requests issued to VFS constitute logical I/O requests. When analyzing VFS, do the following:

- Try to get a picture of the general workload on the system
- Check the frequency of common VFS calls
- Compare the obtained figures with the requests at the physical layer

## Analyzing cache usage

The VFS includes multiple caches to speed up access to frequently used objects. The default behavior in Linux is to complete all write operations in the cache and flush the written data to disk later. Similarly, the kernel also tries to serve the read operations from the cache and shows the page cache hit-and-miss statistics.

The cachestat tool can be used to display statistics for the page cache hit-and-miss ratios:

```
[root@linuxbox ~]# cachestat
   HITS  MISSES  DIRTIES  HITRATIO  BUFFERS_MB  CACHED_MB
 153703    1415      412    99.09%          11      79997
 132378     874      428    99.34%          11      80002
 127758     613     1971    99.52%          11      80007
 128755    1011      828    99.22%          11      80013
 117550     659      801    99.44%          11      80017
 125778     701      547    99.45%          11      80021
 133790     744      595    99.45%          11      80026
 130613     809      732    99.38%          11      80031
  73892    5864      649    92.65%          11      80055
  53714    9304      715    85.24%          11      80081
  55770    8362      435    86.96%          11      80115
```

Figure 10.8 – Using cachestat

From the preceding figure, we can see an excellent cache hit ratio, sometimes even close to 100 percent. This indicates that the kernel is able to satisfy the application's I/O requests from the memory. The higher the percentage of cache hits, the better the performance gains for the application.

Similarly, the `cachetop` tool provides process-wise statistics for the cache hits and misses. The output is displayed on an interactive interface like the `top` command:

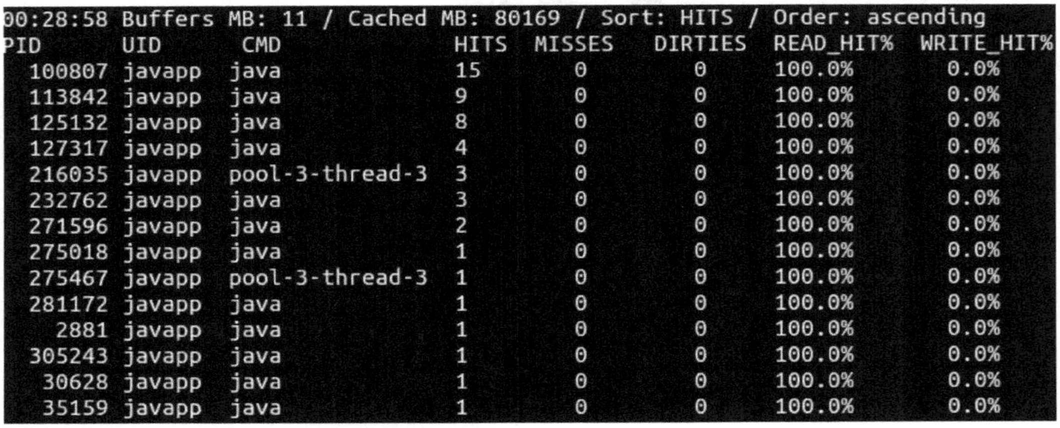

```
00:28:58 Buffers MB: 11 / Cached MB: 80169 / Sort: HITS / Order: ascending
PID       UID    CMD            HITS  MISSES  DIRTIES  READ_HIT%  WRITE_HIT%
   100807 javapp java           15    0       0        100.0%     0.0%
   113842 javapp java           9     0       0        100.0%     0.0%
   125132 javapp java           8     0       0        100.0%     0.0%
   127317 javapp java           4     0       0        100.0%     0.0%
   216035 javapp pool-3-thread-3 3    0       0        100.0%     0.0%
   232762 javapp java           3     0       0        100.0%     0.0%
   271596 javapp java           2     0       0        100.0%     0.0%
   275018 javapp java           1     0       0        100.0%     0.0%
   275467 javapp pool-3-thread-3 1    0       0        100.0%     0.0%
   281172 javapp java           1     0       0        100.0%     0.0%
     2881 javapp java           1     0       0        100.0%     0.0%
   305243 javapp java           1     0       0        100.0%     0.0%
    30628 javapp java           1     0       0        100.0%     0.0%
    35159 javapp java           1     0       0        100.0%     0.0%
```

Figure 10.9 – Using cachetop

When using these tools to analyze cache usage, do the following:

- Look for the hits-and-misses ratio to understand what percentage of requests are being served from memory
- If the ratio is on the lower side, the application or operating system parameters might need to be tuned

## Analyzing filesystems

Although there aren't many tools that can trace filesystem-level operations, BCC offers a few excellent scripts to observe filesystems. Two scripts, `ext4slower` and `xfsslower`, are used to analyze the slow operations on the two most frequently used filesystems, Ext4 and XFS.

The output for both tools, `ext4slower` and `xfsslower`, is identical. By default, both tools print operations that take more than 10 ms to complete, but you can change that by passing the duration value as a parameter. Both tools can also be attached to a specific process:

```
[root@linuxbox ~]# ext4slower
Tracing ext4 operations slower than 10 ms
TIME      COMM PID    T BYTES   OFF_KB      LAT(ms) FILENAME
00:46:55 java 52221  R 4096    0           13.44 appHelper.class
00:46:55 java 52221  R 4096    0           11.88 package.properties
00:46:57 java 52221  W 4096    848         13.97 MyServicesAPI.jar
00:46:57 java 52221  W 4096    1872        12.10 MyServicesAPI.jar
00:46:57 java 52221  W 4096    336         12.61 ant-1.10.11.jar
00:46:57 java 52221  W 4096    848         13.45 aspectjtools.jar
00:46:57 java 52221  R 4096    1872        17.55 aspectjtools.jar
00:46:57 java 52221  R 4096    3920        11.26 aspectjtools.jar
00:46:58 java 52221  R 4096    848         11.04 batik-all-1.14.jar
00:46:58 java 52221  R 4096    848         11.74 bcprov-jdk15on-1.69.jar
00:46:58 java 52221  R 4096    1872        11.05 bcprov-jdk15on-1.69.jar
```

Figure 10.10 – Tracing the slow Ext4 operations

The **T** column shows the type of operation, which can be **R** for read, **W** for write, and **O** for open. The **BYTES** column shows the size of the I/O in bytes, while the **OFF_KB** column shows the file offset for the I/O, in KB. The most important values come from the **LAT(ms)** column, which shows the duration of an I/O request, measured from when it was issued by VFS to the filesystem to when it was completed. This is a fairly accurate measure of the latency endured by an application while performing filesystem I/O.

Another couple of tools included in this set are xfsdist and ext4dist. Both tools show the same information, just for different filesystems – that is, XFS and Ext4, respectively. These tools summarize the time spent while performing common filesystem operations and provide a breakdown of the distribution of the experienced latencies as histograms. Both these tools can be attached to specific processes:

```
[root@linuxbox ~]# xfsdist
Tracing XFS operation latency... Hit Ctrl-C to end.
^C

operation = 'read'
     usecs               : count    distribution
         0 -> 1          : 155503   |***                                     |
         2 -> 3          : 890601   |****************************************|
         4 -> 7          : 87018    |*                                       |
         8 -> 15         : 6159     |                                        |
        16 -> 31         : 5937     |                                        |
        32 -> 63         : 883      |                                        |
        64 -> 127        : 442      |                                        |

operation = 'open'
     usecs               : count    distribution
         0 -> 1          : 234359   |****************************************|
         2 -> 3          : 1261     |                                        |
         4 -> 7          : 203      |                                        |
         8 -> 15         : 305      |                                        |
        16 -> 31         : 80       |                                        |
        32 -> 63         : 11       |                                        |
        64 -> 127        : 2        |                                        |
```

Figure 10.11 – Using xfsdist

When using filesystem-specific tools, remember the following:

- The `ext4dist/xfsdist` tools can help to establish a baseline – that is, whether a workload is read- or write-oriented.

- The two `ext4slower/xfsslower` scripts are extremely effective in determining the actual latency experienced by a process when performing filesystem I/O. When running these, check the latency column to determine the amount of delay being endured by the application.

## Analyzing block I/O

As we saw in *Chapter 9*, the standard disk analysis tools such as `iostat` provide information pertaining to the number of bytes read and written per second, disk utilization, and request queues associated with specific devices. These metrics are averaged out over a period of time and do not offer insights on a per-I/O basis. Extracting information about what happened at a specific interval is not possible.

Similar to VFS and filesystems, the BCC also includes several tools that can help to analyze the events happening in the block layer. One of these tools is `biotop`, which is like a `top` command for disks. By default, the `biotop` tool traces the I/O operations on the block device and displays a summary of each process's activity every second. The summary is sorted based on the top disk consumers in terms of throughput, measured in KB. The process ID and name displayed in the summary represent the time when an I/O operation was initially created, which helps to identify the responsible process:

```
[root@linuxbox ~]# biotop
8:05:10 loadavg: 3.20 5.22 5.59 2/344 194647

PID    COMM            D MAJ MIN DISK      I/O  Kbytes   AVGms
194579 kworker/u8:2    W 8   0   sda       210  155644.0  63.93
194619 cp              R 8   0   sda       135  106656.0 124.26
194637 cp              R 8   0   sda       113  97272.0  153.71
194625 cp              R 8   0   sda       116  93696.0  132.95
194631 cp              R 8   0   sda       116  87296.0  162.71
142538 kworker/u562:1  R 8   0   sda       11   352.0    102.47
```

Figure 10.12 – Using biotop

Another BCC tool to analyze the block layer is `biolatency`. As the name suggests, `biolatency` traces block device I/O and prints a histogram that shows the distribution of I/O latency:

```
[root@linuxbox ~]# biolatency
Tracing block device I/O... Hit Ctrl-C to end.
^C
                usecs           : count   distribution
           0 -> 1              : 0       |                                        |
           2 -> 3              : 0       |                                        |
           4 -> 7              : 0       |                                        |
           8 -> 15             : 0       |                                        |
          16 -> 31             : 0       |                                        |
          32 -> 63             : 13      |                                        |
          64 -> 127            : 913     |**                                      |
         128 -> 255            : 7089    |****************************************|
         256 -> 511            : 4138    |***********                             |
```

Figure 10.13 – Using biolatency

As evident from the preceding output, the bulk of I/O requests took 128–255 microseconds to complete. Depending on the workload, these figures can be much higher.

The `biosnoop` tool from the BCC traces block device I/O and prints the details, including the process that initiated the request:

```
[root@linuxbox ~]# biosnoop
TIME(s)     COMM            PID     DISK   T SECTOR     BYTES  LAT(ms)
0.000000    kworker/u561:2  166418  sda    W 480523072  24576    0.05
0.528009    kworker/u562:0  166544  sda    W 88084480   8192     0.09
0.528023    kworker/u562:0  166544  sda    W 92278792   4096     0.10
0.528029    kworker/u562:0  166544  sda    W 92278912   4096     0.10
0.528032    kworker/u562:0  166544  sda    W 92279088   4096     0.10
0.528034    kworker/u562:0  166544  sda    W 92279120   4096     0.10
0.528036    kworker/u562:0  166544  sda    W 92279136   8192     0.10
0.528039    kworker/u562:0  166544  sda    W 92279160   4096     0.10
0.528042    kworker/u562:0  166544  sda    W 92279568   4096     0.10
0.528044    kworker/u562:0  166544  sda    W 92345032   4096     0.10
```

Figure 10.14 – Using biosnoop

The output from `biosnoop` includes the latency from the time the request was issued to the device to its completion. The `biosnoop` output can be used to identify the process responsible for excessive writes to a disk.

One final tool that I want to mention is `bitesize`, which is used to characterize the distribution of block device I/O sizes:

```
[root@linuxbox ~]# bitesize
Tracing block I/O... Hit Ctrl-C to end.
^C
Process Name = kworker/10:1H
    Kbytes               : count      distribution
        0 -> 1           : 5161       |****************************************|

Process Name = javaMyApp
    Kbytes               : count      distribution
       16 -> 32          : 2877       |****************************************|

Process Name = mysql
    Kbytes               : count      distribution
        4 -> 8           : 3064       |****************************************|
```

Figure 10.15 – Using bitesize

As shown in the preceding output, the **javaMyApp** process (a simple Java-based application) generates requests between 16–32 KB, whereas **mysql** uses the 4–8 KB range.

When analyzing the block layer, remember the following:

- To get a top-end view of disk activity in your system, use `biotop`.

- To trace application I/O sizes, use `bitesize`. If the application workload is sequential, then using larger block sizes might result in better performance.

- To observe block device latencies, use `biolatency`. This will summarize the time ranges for the block I/O requests. If you see higher values, then further digging is required.

- To check further, use `biosnoop`. To find out the time spent between the creation of an I/O request and being issued to a device, use the `-Q` flag with `biosnoop`.

## Summarizing the tools

The following table shows a summary of the tools that can be used to analyze the events in different layers:

| Layer | Analysis tools |
|---|---|
| Application | Application-specific tools |
| System call interface | `strace` and `syscount` (BCC) |
| VFS | `vfsstat`, `vfscount`, and `funccount` |
| Cache | `slabtop`, `cachestat`, `cachetop`, and `dcstat`, `dcsnoop` |
| Filesystems | `ext4slower`, `xfsslower`, `ext4dist`, `xfsdist`, `filetop`, `fileslower`, `stackcount`, `funccount`, `nfsslower`, and `nfsdist` |
| Block layer | `biolatency`, `biosnoop`, `biotop`, `bitesize`, and `blktrace` |
| Disk | `iostat`, `iotop`, `systemtap`, `vmstat`, and PCP |

Table 10.3 – A summary of tools

Note that the tools are not limited to the ones mentioned in the table. The BCC toolset alone includes several other tools that can be used for performance analysis. Further, there are multiple arguments that can be passed to each tool to get a more meaningful output. Considering the multiple layers involved in the hierarchy, diagnosing I/O performance issues is a complex task, and as with any other troubleshooting scenario, it will require the involvement of multiple teams.

## Summary

In this chapter, we resumed our performance analysis and extended it to the higher layers in the I/O stack. Most of the time, analyzing higher layers is skipped, and focus is solely kept on the physical layer. However, for time-sensitive applications, we need to broaden our approach and look for the potential source of delays in application response times.

We started this chapter by explaining the different sources of delays that can be observed by an application when reading from or writing to a filesystem. Filesystems operations go beyond the I/O requests initiated by an application. In addition to application I/O requests, a filesystem can spend time on tasks such as performing metadata updates, journaling, or flushing existing cached data to disks. All these result in extra operations, which incur extra I/O operations. The tools discussed in *Chapter 9* were centered around disks and didn't offer much visibility into the events happening in the VFS and the block layer. The BCC offers a rich set of scripts that can trace the events in the kernel and give us insight into individual I/O requests.

In the next chapter, we'll take our analysis further and learn the different tweaks that we can apply at different levels in the I/O hierarchy, improving performance.

# 11

# Tuning the I/O Stack

Well, here we are at the end of our journey. Just because you are reading the introduction of the final chapter does not mean that you've read through the entire book, but I'll take my chances. If you've indeed followed us along, then I hope your journey was worth it and has left you yearning for more.

Getting back to brass tacks, the previous two chapters centered on the performance analysis of the I/O stack. *Chapter 9* focused on the most common disk metrics and the tools that can help us to identify performance bottlenecks in physical disks. In any performance analysis, the physical disks come under far more scrutiny than any other layer, which can sometimes be misleading. Therefore, in *Chapter 10*, we saw how we can investigate the higher layers in the I/O stack, such as filesystems and the block layer.

This brings us to the next logical step in our quest. Once we've identified the elements plaguing our environment, what steps can we take to mitigate those limitations? It is important to have specific goals for the desired tuning outcomes because, at the end of the day, performance tuning is a trade-off between choices. For instance, tuning the system for low latency may reduce its overall throughput. A performance baseline should first be determined, and any tweaks or adjustments should be carried out in small sets. This chapter will deal with the different tweaks that can be applied to improve I/O performance.

Here's an outline of what follows:

- How memory usage affects I/O
- Tuning the memory subsystem
- Tuning the filesystem
- Choosing the right scheduler

# Technical requirements

The material presented in this chapter builds on the concepts discussed in preceding chapters. If you've followed along and have become familiar with the functions of each layer in the disk I/O hierarchy, you'll find this chapter much easier to follow. If you have a prior understanding of memory management in Linux, that will be a huge plus.

The commands and examples presented in this chapter are distribution-agnostic and can be run on any Linux operating system, such as Debian, Ubuntu, Red Hat, or Fedora. There are quite a few references to the kernel source code. If you want to download the kernel source, you can download it from `https://www.kernel.org`.

# How memory usage affects I/O

As we've seen, VFS serves as an entry point for our I/O requests and includes different types of caches, the most important of which is the page cache. The purpose of page cache is to improve I/O performance and minimize the expense of I/O as generated by swapping and file system operations, thus avoiding unnecessary trips to the underlying physical disks. Although we haven't explored it in these pages, it is important to have an idea about how the kernel goes about managing its memory management subsystem. The memory management subsystem is also referred to as the **virtual memory manager** (**VMM**). Some of the responsibilities of the virtual memory manager include the following:

- Managing the allocation of physical memory for all the user space and kernel space applications
- Implementation of virtual memory and demand paging
- The mapping of files into processes address space
- Freeing up memory in case of shortage, either by pruning or swapping caches

As it's often said, *the best I/O is the one that is avoided*. The kernel follows this approach and makes generous allocation of the free memory, filling it up with the different types of caches. The greater the amount of free memory available, the more effective the caching mechanism. This all works well for general use cases, where applications perform small-scale requests and there is a relative amount of page caches available:

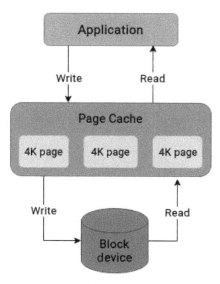

Figure 11.1 – A page cache can speed up I/O performance

Conversely, if memory is scarcely available, not only will the caches be pruned regularly but the data might also get swapped out to disk, which will ultimately hurt performance. The kernel works under the **temporal locality principle**, meaning that the recently accessed blocks of data are more likely to be accessed again. This is generally good for most cases. It could take a few milliseconds to read data from a random part of the disk, whereas accessing that same data from memory if it is cached only takes a few nanoseconds. Therefore, any request that can be readily served from the page cache minimizes the cost of an I/O operation.

## Tuning the memory subsystem

It's a bit strange that how Linux deals with memory can have a major say in disk performance memory. As already explained, the default behavior of the kernel works well in most cases. However, as they say, an excess of everything is bad. Frequent caching can result in a few problematic scenarios:

- When the kernel has accumulated a large amount of data in the page cache and eventually starts to flush that data onto disk, the disk will remain busy for quite some time because of the excessive write operations. This can adversely affect the overall I/O performance and increase disk response times.

- The kernel does not have a sense of the criticality of the data in the page cache. Hence, it does not distinguish between *important* and *unimportant* I/O. The kernel picks whichever block of data it deems appropriate and schedules it for a write or read operation. For instance, if an application performs both background and foreground I/O operations, then usually, the priority of the foreground operations should be higher. However, I/O belonging to background tasks can overwhelm foreground tasks.

The cache provided by the kernel usually enables applications to obtain better performance when reading and writing data, but the algorithms used by the page cache are not designed for a particular application; they are designed to be general-purpose. In most cases, this default behavior would work just fine, but in some cases, this can backfire. For some self-caching applications, such as database management systems, this approach might not offer the best results. Applications such as databases have a better understanding of the way data is organized internally. Hence, these systems prefer to have their own caching mechanism to improve read and write performance.

## Using direct I/O

If data is cached directly at the application level, then moving data from disk to the page cache and back to the application's cache will constitute a significant overhead, resulting in more CPU and memory usage. In such scenarios, it might be desirable to bypass the kernel's page cache altogether and leave the responsibility of caching to the application. This is known as **direct I/O**.

Using direct I/O, all the file reads and writes go directly from the application to the storage device, bypassing the kernel's page cache. The **Unix filesystem** (**UFS**) filesystem (not supported in Linux) includes direct I/O as filesystem parameters, which can be specified while mounting the filesystem. In Linux, direct I/O is not a filesystem parameter, nor is there a command to enable it. Instead, it is the responsibility of an application to initiate direct I/O. The application can invoke direct I/O by using the O_DIRECT flag on a system call, such as open (). The O_DIRECT flag is only a status flag (represented by DIR), which is passed by the application while opening or creating a file so that it can go around the page cache of the kernel:

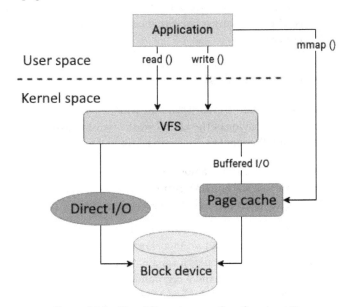

Figure 11.2 – The different ways of performing I/O

It doesn't make sense to use direct I/O for regular applications, as it can cause performance deterioration. However, for self-caching applications, it can offer significant gains. The recommended method is to check the status of direct I/O via an application. However, if you want to check via the command line, use the `lsof` command to check the flags through which a file is opened.

```
[root@linuxbox ~]# lsof +fg /db2/ids_space/test/logdbs_mcp_test
COMMAND    PID    USER    FD    TYPE    FILE-FLAG      DEVICE   SIZE/OFF  NODE      NAME
oninit    9456    root    1w    REG     RW,SYN,DIR,LG  120,240  3019654   12084979  /db2/ids_space/test/logdbs_mcp_test
oninit    9456    root    2w    REG     RW,SYN,DIR,LG  120,240  3019654   12084979  /db2/ids_space/test/logdbs_mcp_test
oninit    9459    root    1w    REG     RW,SYN,DIR,LG  120,240  3019654   12084979  /db2/ids_space/test/logdbs_mcp_test
```

Figure 11.3 – Checking direct I/O

For files opened by the application through the `O_DIRECT` flag, the **FILE-FLAG** column of the output will include the **DIR** flag.

The performance gains from direct I/O come from avoiding the CPU cost of copying data from disk into the page cache, and from steering clear of the double buffering, once in the application and once in the filesystem.

## Controlling the write-back frequency

As already explained, caching has its advantages, as it accelerates many accesses to files. Once most of the free memory has been occupied by the cache, the kernel has to make a decision on how to free memory in order to entertain incoming I/O operations. Using the **Least Recently Used** (**LRU**) approach, the kernel does two things – it evicts old data from the page cache and even offloads some of it to the swap area, in order to make room for incoming requests.

Again, it all comes down to the specifics. The default approach is good enough, and that is exactly how the kernel should go about making room for incoming data. However, consider the following scenarios:

- What if the data currently in the cache won't be accessed again in the future? This is true for most backup operations. A backup operation will read and write a lot of data from the disk, which will be cached by the kernel. However, it is unlikely that this data, which is present in the cache, will be accessed in the near future. However, the kernel will keep this data in the cache and might evict the older pages, which had a greater probability of being accessed again.

- Swapping data to disk will generate a lot of disk I/O, which won't be good for performance.

- When a large amount of data has been cached, a system crash can result in a major loss of data. This can be a significant concern if data is of a sensitive nature.

It's not possible to disable the page cache. Even if there was, it's not something that should be done. There are, however, a number of parameters that can be tweaked to control its behavior. As shown here, there are several parameters that can be controlled through the `sysctl` interface:

```
[root@linuxbox ~]# sysctl -a | grep dirty
vm.dirty_background_ratio = 10
```

```
vm.dirty_background_bytes = 0
vm.dirty_ratio = 20
vm.dirty_bytes = 0
vm.dirty_writeback_centisecs = 500
vm.dirty_expire_centisecs = 3000
```

Let us look at them in detail:

- `vm.dirty_background_ratio`: The write-back flusher threads initiate the flushing of dirty pages to disk when the percentage of dirty pages in the cache surpasses a certain threshold. Prior to this threshold, no pages are written to the disk. Once flushing begins, it occurs in the background without causing any disturbance to the foreground processes.

- `vm.dirty_ratio`: This refers to the threshold of system memory utilization, beyond which the writing process gets blocked and dirty pages are written out to the disk.

For large memory systems, hundreds of GB of data can be flushed from the page cache to disk, which will cause noticeable delays and adversely affect not only the disk's performance but also overall system performance. In such cases, lowering these values might be helpful, as data will be flushed to disk on a regular basis, avoiding the write storm.

You can check the current values of these parameters using `sysctl` – for instance, if the values are as follows:

```
vm.dirty_background_ratio=10
vm.dirty_ratio=20
```

Think of `vm.dirty_ratio` as the upper limit. Using these previously mentioned values means that when the percentage of dirty pages in the cache reaches 10%, the background threads are triggered to write them to the disk. However, when the total number of dirty pages in the cache exceeds 20%, all writes are blocked until a portion of the dirty pages are written to the disk. These two parameters have the following two counterparts:

- `vm.dirty_background_bytes`: This denotes the amount of dirty memory, expressed in bytes, that triggers the background flusher threads to initiate writing back to the disk. This is the counterpart of `vm.dirty_background_ratio`, and only one of them can be configured. The value can be defined either as a percentage or a precise number of bytes.

- `vm.dirty_bytes`: This is the amount of dirty memory, expressed in bytes, that results in the writing process getting blocked and writing out the dirty pages to the disk. This controls the same tunable as `vm.dirty_ratio`, and only one of them can be set.

- `vm.dirty_expire_centisecs`: This indicates how long something can be in the cache before it needs to be written. This tunable specifies the age at which the dirty data is deemed suitable for write-back by the flusher threads. The time duration is measured in hundredths of a second.

To summarize, the default behavior of the kernel's page cache works well most of the time, and usually, it won't require any tweaking. However, for intelligent applications such as large-scale databases, the frequent caching of data can become a hurdle. Fortunately, there are a few workarounds available. Such applications can be configured to use direct I/O, which will bypass the page cache. The kernel also offers several parameters that can be used to tweak the behavior of the page cache. However, it is important to note that changing these values can result in increased I/O traffic. Therefore, workload-specific testing should be conducted prior to making changes.

# Tuning the filesystem

As we focus on tuning the different components that can impact I/O performance, we'll try to steer our conversation away from the hardware side of things. Given the advancements in hardware, upgrading the memory, compute, network, and storage apparatus is bound to add at least some level of performance gains. However, most of the time, the magnitude of those gains will be limited. You need a well-designed and configured software stack to take advantage of that hardware. As we're not focusing on a particular type of application, we'll try to present some general tweaks that can be used to fine-tune your I/O. Again, note that the parameters that will be presented here or were discussed earlier require thorough testing and may not offer the same results in different environments.

Coming back to the topic of our discussion, filesystems are responsible for organizing data on disk and are the point of contact for an application to perform I/O. This makes them an ideal candidate for the tuning and troubleshooting process. Some applications explicitly mention the filesystem that should be used for optimal performance. As Linux supports different flavors of filesystems that use different techniques to store user data, some mount options might not be common among filesystems.

## Block size

Filesystems address physical storage in terms of blocks. A **block** is a group of physical sectors and is the fundamental unit of I/O for a filesystem. Each file in the filesystem will occupy at least one block, even if the file contains nothing. By default, a block size of 4 KB is used for most filesystems. If the application mostly creates a large number of small-sized files in the filesystem, typically of a few bytes or less than a couple of KB, then it is best to use smaller block sizes than the default value of 4 KB.

Filesystems perform better if applications use the same read and write size as the block size, or use a size that is a multiple of the block size. The block size for a filesystem can only be specified during its creation and cannot be changed afterward. Therefore, the block size needs to be decided before creating the filesystem.

## Filesystem I/O alignment

The concept of I/O alignment is generally overlooked, but this can have a huge impact on the filesystem performance. This is especially true for the complex enterprise storage systems of today, which consist of flash drives that have different page sizes and some form of RAID configuration on top of them.

The I/O alignment for filesystems is concerned with how data is distributed and organized across the filesystem. That's one side of the coin. If the underlying physical storage consists of a striped RAID configuration, the data should be aligned with the underlying storage geometry for optimal performance. For instance, for a RAID device with a 64 K per-disk stripe size and 10 data-bearing disks, the filesystem should be created as follows:

- For XFS, it should be the following:

```
mkfs.xfs -f -d su=64k,sw=10   /dev/sdX
```

XFS provides two sets of tunables in this regard. Depending on the specification units that you've set, you can use these:

  - `sunit`: A stripe unit, in 512-byte blocks
  - `swidth`: A stripe width, in 512-byte blocks

  Alternatively, you can use these:

  - `su`: A per-disk stripe unit, in K if suffixed with *k*
  - `sw`: A stripe width, by the number of data disks

- For Ext4, the command would look as follows:

```
mkfs.ext4 -E stride=16,stripe-width=160 /dev/sdX
```

Ext4 also provides a couple of tunables, which can be used as follows:

  - `stride`: The number of filesystem blocks on each data-bearing disk in that stripe
  - `stripe-width`: The total stripe-width in filesystem blocks, equal to (stride) x (the number of data-bearing disks)

## LVM I/O alignment

Every abstraction layer created on top of a RAID device must be aligned to a multiple of `Stripe Width`, plus any required initial alignment offset. This ensures that a read or write request of a single block at the filesystem will not span the RAID stripe boundaries and cause multiple stripes to be read and written at the disk level, adversely affecting performance.

The first physical extent allocated within the physical volume should be aligned to a multiple of the RAID Stripe Width. If the physical volume is created directly on a raw disk, then it should also be offset by any required initial alignment offset. To check where the physical extents start, use the following:

```
pvs -o +pe_start
```

Here, pe_start refers to the first physical extent.

The logical volumes are always allocated a contiguous range of physical extents when possible. If a contiguous range doesn't exist, non-contiguous ranges might be allocated. Since a non-contiguous range of extents can impact performance, there is an option (--contiguous) while creating a logical volume to prevent the non-contiguous allocation of extents.

## Journaling

As explained in *Chapter 3*, the concept of journaling guarantees data consistency and integrity if I/O operations on a filesystem fail due to external events. Any changes that need to be performed on the filesystem are first written to a journal. Once data has been written to a journal, it is then written to the appropriate location on the disk. If there is a system crash, the filesystem replays the journal to see if any operation was left in an incomplete state. This reduces the likelihood that the filesystem will become corrupted if there are any hardware failures.

Apparently, the journaling approach adds extra overhead and can potentially affect filesystem performance. However, given the sequential nature of journal writes, the filesystem performance is not affected. So, it is recommended to keep the filesystem journal enabled to ensure data integrity.

It is, however, recommended to change the mode of the filesystem journal to suit your needs. Most filesystems don't have multiple journaling modes but Ext4 offers a great deal of flexibility in this regard. The Ext4 offers three journaling modes. Among them, the **write-back mode** offers considerably better performance than the ordered and data mode. The write-back mode only journals the metadata and does not follow any order when writing the data and metadata to disk. The **ordered mode** on the other hand follows a strict order and first writes the actual data before the metadata. The **data mode** offers the lowest performance, as it has to write both the data and metadata to a journal, resulting in twice the number of operations.

Another thing that can be done to improve journaling is to use an external journal. The default location for a filesystem journal is on the same block device as the data. If the I/O workload is metadata-intensive and the synchronous metadata writes to the journal must complete successfully before any associated data writes can start, this can result in I/O contention and may impact performance. In such cases, it can be a good idea to use an external device for filesystem journaling. The journal size is typically very small and requires very little storage space. The external journal should ideally be placed on fast physical media with a battery-backed write-back cache.

## Barriers

As mentioned earlier, most filesystems make use of journaling to keep track of changes that have not yet been written to disk. A **write barrier** is a kernel mechanism that guarantees the proper ordering and accurate writing of filesystem metadata onto persistent storage, even if storage devices with unstable write caches lose power. Write barriers enforce proper on-disk ordering of journal commits by forcing the storage device to flush its cache at certain intervals. This makes volatile write caches safe to use, but it can incur some performance deficit. If the storage device cache is battery-backed, disabling filesystem barriers may offer some performance improvement.

## Timestamps

The kernel records information about when files were created (`ctime`) and last modified (`mtime`) ,as well as when they were last accessed (`atime`). If an application frequently modifies a bunch of files, then their corresponding timestamps will need to be updated every time. Performing these modifications also requires I/O operations, and when there are too many of them, there is a cost associated with them.

To mitigate this, there is a special mount option for filesystems called `noatime`. When a filesystem is mounted with the `noatime` option, reading from the filesystem will not update the file's `atime` information. The `noatime` setting is significant, as it removes the requirement for a system to perform writes to the filesystem for files that are only read. This can lead to noticeable performance improvements, since write operations can be expensive.

## Read-ahead

The read-ahead functionality in filesystems can enhance file access performance by proactively fetching data that is expected to be required soon and storing it in the page cache, which allows for faster access compared to retrieving the data from the disk. A higher read-ahead value indicates that the system will prefetch data further ahead of the current read position. This is especially true for sequential workloads.

## Discarding unused blocks

As we explained in *Chapter 8*, in SSDs, a write operation can be done at the page level, but the erase operation always affects entire blocks. As a result, writing data to SSDs is very fast as long as empty pages can be used. However, once previously written pages need to be overwritten, the writes slow down considerably, impacting performance. The `trim` command tells the SSD to discard the blocks that are no longer needed and can be deleted. The filesystem is the only component in the I/O stack that knows the parts of the SSD that should be trimmed. Most filesystems offer mount parameters that implement this feature.

To summarize, filesystems, to a certain extent, map logical addresses to physical addresses. When an application writes data, the filesystem decides how to distribute writes properly in order to make the best use of the underlying physical storage. This makes filesystems a very important layer when it comes to performance tuning. Most of the changes in filesystems cannot be done on the fly; they're either performed during filesystem creation or require unmounting and remounting the filesystem. So, the decision regarding the choice of filesystem parameters should be made in advance, as changing things afterward can be a disruptive activity.

## Choosing the right scheduler

The sole purpose of I/O schedulers is to optimize disk access requests. There are some common techniques used by schedulers, such as merging I/O requests that are adjacent on disk. The idea is to avoid frequent trips to the physical storage. Aggregating requests that are situated in close proximity on the disk reduces the frequency of the drive's seeking operations, thereby enhancing the overall response time of disk operations. I/O schedulers aim to optimize throughput by rearranging access requests into sequential order. However, this strategy may cause some I/O requests to wait for an extended time, resulting in latency problems in certain situations. I/O schedulers strive to achieve a balance between maximizing throughput and distributing I/O requests equitably among all processes. As with all other things, Linux has a variety of I/O schedulers available. Each has its own set of strengths:

| Use case | Recommended I/O scheduler |
| --- | --- |
| Desktop GUI, interactive applications, and soft real-time applications, such as audio and video players | Budget Fair Queuing (BFQ), as it guarantees good system responsiveness and a low latency for time-sensitive applications |
| Traditional mechanical drives | BFQ or Multiqueue (MQ)-Deadline – both are considered suitable for slower drives. Kyber/none are biased in favor of faster disks. |
| High-performing SSDs and NVMe drives as local storage | Preferable none, but Kyber might also be a good alternative in some cases |
| Enterprise storage arrays | None, as most storage arrays have built-in logic to schedule I/Os more efficiently |
| Virtualized environments | MQ-Deadline is a good option. If the hypervisor layer does its own I/O scheduling, then using the none scheduler might be beneficial, |

Table 11.1 – Some use cases for I/O schedulers

The good thing is that an I/O scheduler can be changed on the fly. It is also possible to use a different I/O scheduler for every storage device on the system. A good starting point to select or fine-tune an I/O scheduler is to determine the system's purpose or role. It is generally accepted that there is no single I/O scheduler that can meet all of a system's diverse I/O demands.

# Summary

After having spent two chapters trying to diagnose and analyze the performance of different layers in the I/O stack, this chapter focused on the performance-tuning aspect of the I/O stack. Throughout this book, we've familiarized ourselves with the multi-level hierarchy of the I/O stack and built an understanding of the components that can impact the overall I/O performance.

We started this chapter by briefly going through the functions of the memory subsystem and how it can impact the I/O performance of a system. As all write operations are, by default, first performed in the page cache, the way the page cache is configured to behave can have a major say in an application's I/O performance. We also explained the concept of direct I/O and defined some of the different parameters that can be used to tweak the write-back cache.

We also looked at the different tuning options when it comes to filesystems. Filesystems offer different mount options that can be changed to reduce some I/O overhead. Additionally, the filesystem block size, its geometry, and I/O alignment in terms of the underlying RAID configuration can also impact performance. Finally, we explained some use cases of the different scheduling flavors in Linux.

I guess that's a wrap! I sincerely hope that this book took you on an enlightening exploration of the intricate layers comprising the Linux kernel's storage stack. Starting with an introduction to VFS in *Chapter 1*, we tried to navigate the complex terrain of storage architecture. Each chapter delves deeper into the intricacies of the Linux storage stack, exploring topics such as VFS data structures, filesystems, the role of the block layer, multi-queue and device-mapper frameworks, I/O scheduling, the SCSI subsystem, physical storage hardware, and its performance tuning and analysis. Our goal was to prioritize the conceptual side of things and examine the flow of disk I/O activity, which is why we've not dived too much into general storage administration tasks.

As we bring our adventure to a close, I hope you've gained a comprehensive understanding of the Linux storage stack, its major components, and their interactions, and now possess the necessary knowledge and skills to make informed decisions, analyze, troubleshoot, and optimize storage performance in Linux environments.

# Index

www.packtpub.com

Subscribe to our online digital library for full access to over 7,000 books and videos, as well as industry leading tools to help you plan your personal development and advance your career. For more information, please visit our website.

## Why subscribe?

- Spend less time learning and more time coding with practical eBooks and Videos from over 4,000 industry professionals

- Improve your learning with Skill Plans built especially for you

- Get a free eBook or video every month

- Fully searchable for easy access to vital information

- Copy and paste, print, and bookmark content

Did you know that Packt offers eBook versions of every book published, with PDF and ePub files available? You can upgrade to the eBook version at packtpub.com and as a print book customer, you are entitled to a discount on the eBook copy. Get in touch with us at customercare@packtpub. com for more details.

At www.packtpub.com, you can also read a collection of free technical articles, sign up for a range of free newsletters, and receive exclusive discounts and offers on Packt books and eBooks.

# Other Books You May Enjoy

If you enjoyed this book, you may be interested in these other books by Packt:

**The Ultimate Kali Linux Book - Second Edition**

Glen D. Singh

ISBN: 9781801818933

- Explore the fundamentals of ethical hacking
- Understand how to install and configure Kali Linux
- Perform asset and network discovery techniques
- Focus on how to perform vulnerability assessments
- Exploit the trust in Active Directory domain services
- Perform advanced exploitation with Command and Control (C2) techniques
- Implement advanced wireless hacking techniques
- Become well-versed with exploiting vulnerable web applications

**Mastering Linux Security and Hardening - Third Edition**

Donald A. Tevault

ISBN: 9781837630516

- Prevent malicious actors from compromising a production Linux system
- Leverage additional features and capabilities of Linux in this new version
- Use locked-down home directories and strong passwords to create user accounts
- Prevent unauthorized people from breaking into a Linux system
- Configure file and directory permissions to protect sensitive data
- Harden the Secure Shell service in order to prevent break-ins and data loss
- Apply security templates and set up auditing

## Packt is searching for authors like you

If you're interested in becoming an author for Packt, please visit authors.packtpub.com and apply today. We have worked with thousands of developers and tech professionals, just like you, to help them share their insight with the global tech community. You can make a general application, apply for a specific hot topic that we are recruiting an author for, or submit your own idea.

## Share Your Thoughts

Now you've finished *Architecture and Design of the Linux Storage Stack*, we'd love to hear your thoughts! Scan the QR code below to go straight to the Amazon review page for this book and share your feedback or leave a review on the site that you purchased it from.

https://packt.link/r/1837639965

Your review is important to us and the tech community and will help us make sure we're delivering excellent quality content.

# Download a free PDF copy of this book

Thanks for purchasing this book!

Do you like to read on the go but are unable to carry your print books everywhere?

Is your eBook purchase not compatible with the device of your choice?

Don't worry, now with every Packt book you get a DRM-free PDF version of that book at no cost.

Read anywhere, any place, on any device. Search, copy, and paste code from your favorite technical books directly into your application.

The perks don't stop there, you can get exclusive access to discounts, newsletters, and great free content in your inbox daily

Follow these simple steps to get the benefits:

1. Scan the QR code or visit the link below

https://packt.link/free-ebook/9781837639960

2. Submit your proof of purchase

3. That's it! We'll send your free PDF and other benefits to your email directly